Eating Earth

Eating Earth

Environmental Ethics and
Dietary Choice

LISA KEMMERER

OXFORD
UNIVERSITY PRESS

Oxford University Press is a department of the University of Oxford.
It furthers the University's objective of excellence in research, scholarship,
and education by publishing worldwide.

Oxford   New York
Auckland   Cape Town   Dar es Salaam   Hong Kong   Karachi
Kuala Lumpur   Madrid   Melbourne   Mexico City   Nairobi
New Delhi   Shanghai   Taipei   Toronto

With offices in
Argentina   Austria   Brazil   Chile   Czech Republic   France   Greece
Guatemala   Hungary   Italy   Japan   Poland   Portugal   Singapore
South Korea   Switzerland   Thailand   Turkey   Ukraine   Vietnam

Oxford is a registered trademark of Oxford University Press
in the UK and certain other countries.

Published in the United States of America by
Oxford University Press
198 Madison Avenue, New York, NY 10016

CIP data is on file at the Library of Congress
ISBN 978-0-19-939184-4

9 8 7 6 5 4 3 2 1
Printed in the United States of America
on acid-free paper

For Mum and Pops. Thanks for giving me so much freedom in the out-of-doors.

Contents

List of Figures and Slides

Figures

Slides

Acknowledgments

MANY THANKS TO three anonymous Oxford reviewers. Special thanks to Erik Hane, Eswari Marudhu, Ben Saddock, Mathematician Maggie McBride, and to my friend and colleague John Halley for providing feedback on the manuscript, and for taking the lead in creating figures (that only a math expert could manage) to help readers more easily visualize/grasp key ideas presented in this book.

Eating Earth

Introduction

WHEN I WAS in my twenties on a Watson Fellowship that took me to the Tibetan Plateau, I met an Irishman at a low-end restaurant. We had dinner together. Though each of us had met many other people in our travels, there was something magnetic about our connection—I refer to the negative ends of a magnet. We disagreed about pretty much everything. At the time, I would best have been described as a budding philosopher, ethicist, atheist, feminist, and animal liberationist; he was a scientist, mathematician, born-again Christian, and environmentalist. Ideologically we had almost nothing in common, but since we had both been traveling for months with little opportunity to speak English (and even less to engage in meaningful discussions) we spent our days together . . . and argued almost perpetually.

Because we tended to meet at dinner, and because I was a vegetarian, our disagreements usually began over food, then spread to innumerable other areas of discord. Despite the discord, we continued our discussions long after we returned to our respective homes. Yet neither reason nor heartfelt pleas shifted the Irishman to a plant-based diet. He was sympathetic to human moral responsibilities for animal suffering (ever the Christian), but on learning of the cruelty of animal agriculture, he merely shifted to "happy meat" and the eggs of "free range" chickens. Though I feverishly pointed to the horrific transport and dependable adolescent slaughter of grass-fed and "free range" animals, and the absence of any nutritional need for animal products in our diet, my energy was wasted. Ultimately, it was the Irishman's concern for the environment, combined with his predilection for numbers, that altered his dietary choices.

Recently my friend composed an essay for an anthology I was putting together, *about animal advocacy and environmentalism and the search for common ground*. In the process, he applied his math and science skills to calculate the ratio of the mass of wild birds to the ratio of the mass of chickens in the U.K.—1:104. His horror was palpable despite his wry response: "For every ten grams of wild bird, somewhere out there (and close) lies a full kilo of chicken." In that moment he fully *felt* the link between dietary choice and environmental degradation for the first time. Reflecting on this discovery, he wrote: "This was a 'Eureka' moment, a dark one, a moment that has informed everything since."

Had I understood the environmental impact of consuming animal products when we met in Tibet, perhaps I might have had more success in persuading my friend to shift toward toward a plant-based diet two decades earlier. This was also a eureka moment for me, and it is my inspiration for writing *Eating Earth*.

Ethics and Dietary Choice

Even among environmentalists, few things are as habitual and devoid of conscious thought and scrutiny as what we put into our mouths. This is especially true when we are young: We eat what we are served, and for the most part what we are served when young becomes the basis of what we consume as adults. Our diets don't generally change unless or until someone or something (usually new information or relocation) causes us to see certain foods differently, at which time we discover previously overlooked foods, or perhaps food favorites become morally questionable.

Though our dietary choices tend to be inherited and habitual, we make many choices within this given framework. For example, we might choose not to eat crabs or deer or ducks or herring or strawberries—even though everyone around us does so. *Morally speaking, this ability to choose is critical.* If someone engages in an immoral act, but was compelled to do so at gunpoint, this case must be taken separately from ones in which a moral agent chooses his or her actions in the absence of fear or coercion. Moral accountability presupposes a certain ability to *choose* between one action and another. Though diet is largely a matter of culture and tradition, and though finances and access restrict options for many people, it is likely that anyone reading this book has considerable dietary choice.

Who Is "We"?

Even in wealthy nations there are many people who have little choice regarding food consumption.[1] For example, reasonably priced fresh fruits and vegetables are not necessarily available to inner-city residents, many of whom do not own private transportation, which limits where families are likely to shop, especially poor families who must think carefully about the cost of additional bus fares. Nor is it possible for most inner-city individuals to throw a few seeds in the ground and wait for greens to emerge. Similarly, in other regions of the world poorer people tend to have little choice in what they put on the table. Sadly, while most Americans have tremendous choice regarding diet, many people around the world are happy to have something—anything—on the table. Poverty limits food choices, as does isolation. Because choice is critical to moral accountability, *Eating Earth* is written for those who have a reasonable measure of choice regarding diet: *This book is not intended as a criticism of those who have little or no choice in what they consume.*

This book is about environmental ethics and dietary choice. It is therefore written about and for human beings who have various options with regard to what they eat, and can only legitimately speak to those who have access to a variety of foods and who *choose* what they will bring to the table. Those who have little or no choice regarding diet cannot reasonably be held morally accountable in the same way as those who, like myself, choose to be either an omnivore or a vegan, for example. Again, if you are reading this book, it is highly likely that you have a reasonable measure of choice as to what you consume.

Because the largest population of those for whom this book is written (English-reading individuals who have a choice in what they eat) are likely to be in the United States, much of the information in this book focuses on the U.S., but I have often included information from other nations and regions of the world, especially in light of the fact that environmental concerns tend to be international.

---

1. For insights into different foodways in the U.S., please see A. Breeze Harper, ed., *Sistah Vegan: Black Female Vegans Speak on Food, Identity, Health, and Society* (New York: Lantern, 2010). Also, check out the Food Empowerment Project online (http://www.foodispower. org/).

What Do Animal Agriculture, Fisheries, and Hunting Have in Common?

The first two chapters of this book focus on a particular dietary choice: Animal agriculture and fisheries exist only inasmuch as human beings consume animal products. Animal agriculture only exists when people eat chickens, cattle, pigs, turkeys, eggs, yogurt, cheese, and so on. Fisheries only exist when people eat tuna, salmon, pollack, sardines, shrimp, lobsters, octopus, and so on. Similarly, sport hunting only continues because the larger culture accepts hunting as a legitimate form of recreation, *largely because hunted animals are ultimately consumed,* and acquiring food in this way tends to be viewed as an acceptable enterprise. It is empowering to recognize the complete dependence of animal agriculture, fisheries, and hunting on human dietary choices and on our view of other animals as eatable. Animal agriculture, fisheries, and hunting exist due to human consumption patterns, and their continued existence depends wholly on continuing these consumption patterns. Collectively, our consumption patterns determine whether or not this approach to other animals continues or fades into history.

Why Focus on the Environment?

The Irishman and I have stayed in touch across time and miles. We occasionally get together to go hiking, and to continue working out ongoing disagreements (animal ethics, environment, and religion are still our favorite subjects, though they are now much less points of disagreement than topics of ongoing exploration). The Irishman's recent "eureka moment" has brought home the connection between dietary choice and environmental degradation for both of us. Whether you are a mainstream environmentalists, an active member of the Earth Liberation Front, an ecofeminist, or a deep ecologist, if you care about the health of this planet or the future of humanity, and if you have access to a variety of affordable food alternatives, this book is for you.

1

Farming Facts

Livestock activities have significant impact on virtually all aspects of the environment, including air and climate change, land and soil, water and biodiversity. . . . The overall impact of livestock on the environment is enormous.
—STEINFELD ET AL.

CHEAP MEAT, DAIRY, and eggs are an illusion—we pay for each with depleted forests, polluted freshwater, soil degradation, and climate change. Diet is the most critical decision we make with regard to our environmental footprint—and what we eat is a choice that most of us make every day, several times a day.

Pollution: Greenhouse Gas Emissions and Water

Dietary choice contributes powerfully to greenhouse gas emissions (GHGE) and water pollution.

GHGE

Animal agriculture is responsible for an unnerving quantity of greenhouse gas emissions. Eating animal products—yogurt, ice cream, bacon, chicken salad, beef stroganoff, or cheese omelets—greatly increases an individual's contribution to carbon dioxide, methane, and nitrous oxide emissions. Collectively, dietary choice contributes to a classic "tragedy of the commons."

Fossil Fuels

Much of the atmosphere's carbon dioxide (CO_2) is absorbed by the earth's oceans and plants, but a large proportion lingers in the

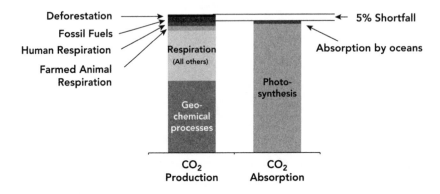

More CO_2 is currently released than can be absorbed by photosynthesis, causing a steady buildup (> 41 million tons per day) of CO_2 in the atmosphere. The above two towers must be of equal height for environmental health and stability — we must reduce CO_2 emissions to make this happen.

FIGURE 1.1 Deadly Imbalance of Carbon Dioxide. Annual production and absorption of CO2.
More CO_2 is currently released than can be absorbed by photosynthesis, causing a steady buildup (> 41 million tons per day) of CO_2 in the atmosphere. The above two columns must be of equal height for environmental health and stability—we must reduce CO_2 emissions to make this happen.

Quantitative graphs, tables and quotations in text are based on the most reliable data currently available. However, some figures have some variability depending on sources, reflecting the current state of knowledge in some of these areas. Every effort has been made to harmonize such figures.

atmosphere—unable to be absorbed by plants or oceans ("Effects").[1] Plants are not harmed by this process, but the current overabundance of carbon dioxide in the atmosphere causes acidification of the earth's oceans. As a result of anthropogenic carbon dioxide emissions, the "acidity of the world's ocean may increase by around 170% by the end of the century," altering ocean ecosystems, and likely creating an ocean environment that is inhospitable for many life forms (Figure 1.1) ("Expert Assessment").

Burning petroleum also leads to wars that devastate human communities and annihilate landscapes and wildlife—including endangered

---

1. Statistics always contain a margin of error. It is critical to note sources of funding for research when dealing with statistics surrounding the powerful animal agriculture industry (Big Ag). Dependable information on the subject of diet cannot come from moneyed interests—from research financed by animal industries or their associates, or from individuals in any way affiliated with such industries and their interests. Furthermore, the plethora of information and the overwhelming evidence offered in this book—and others, such as the U.N. publication *Livestock's Long Shadow*—cannot be discounted merely by reference to possible statistical margins of error.

species and their vital habitats. Additionally, our consumption of petroleum is linked with oil spills that ravage landscapes, shorelines, and ocean habitat. Oil pipelines run through remote, fragile areas—every oil tanker represents not just the possibility but the probability of an oil spill. As reserves diminish, our quest for fossil fuels is increasingly environmentally devastating: Canada's vast reserves of tar sands oil—though extracted, transported, and burned only with enormous costs to the environment—are next in line for extraction.

s1.1 GHGE – The Cost of Consuming Animal Products

Consuming animal products creates ten times more fossil fuel emission *per calorie* than does consuming plant foods directly (Oppenlander 18). (This is the most remarkable given that plant foods are not generally as calorically dense as animal foods.) Ranching is the greatest GHGE offender. The United States and Brazil are the world's leading beef producers. The U.S. consumed nearly 26 billion pounds (12 billion kg) of beef in 2011, supporting an industry with a retail value of almost $80 billion, and exporting 3 billion pounds (1.3 billion kg) of beef, rounding up more than $5 billion ("Cattle and Beef"). All this even though:

- Producing just one protein calorie in feedlot bovines requires nearly 80 calories of fossil fuels, while one protein calorie from soybeans requires just 2 calories of fossil fuels (Schwartz 86).
- One serving of beef creates the atmospheric warming potential of 80 pounds (36 kg) of carbon dioxide—as much as driving an ordinary car three hours to travel 155 miles (250 km) (Fanelli).

- Industrially produced flesh has an energy input-to-output ratio of 35:1 (Cassuto 4).

With regard to diet and the environment, the most important statistic to remember is that *70 percent of U.S. grains and 60 percent of EU grains are fed to farmed animals* (Figure 1.2) (Oppenlander 12; Steinfeld et al. 272). Every year in the U.S., bovines consume roughly 110 billion pounds (50 billion kg) of grain: Producing one pound (0.45 kg) of beef requires an estimated 16 pounds (7.3 kg) of grain (Dawn 280). Farmed animals consume more than 635 million metric tons of grain each year worldwide ("Top 5"). (One metric ton, or tonne, is 1,000 kg; one ton—a U.S. measurement sometimes called a short ton—is 2,000 pounds, or 907 kg. I have used metric tons throughout, except when inside quotations.) Feeding grain to bovines is environmentally costly and foolish: Humans create as much as 37 million tons of CO_2 just to produce the chemical fertilizers that are poured onto feed crops ("Greenhouse Gas Emissions"). Fossil fuels are burned to prepare the land, plant the crops, fertilize the soil, weed and cull crops, and harvest and transport seeds, fertilizer, equipment, green waste, and grains. Additionally, fossil fuels are burned at every facility that breeds, feeds, and maintains bovines (or other farmed animals), as well as for slaughter, processing, and transportation.

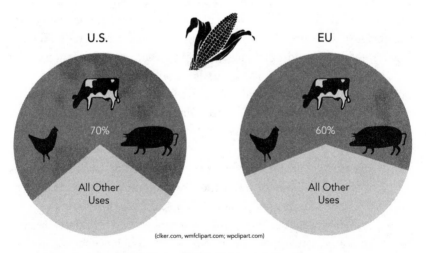

(clker.com, wmfclipart.com; wpclipart.com)

FIGURE 1.2 Percentage of Grains Fed to Farmed Animals

- Each calf at a feedlot consumes a *minimum* of 780 lbs (354 kg) of grain before slaughter.

- Each calf generally consumes more than a ton of grain in just 4 months before slaughter.

- Lactating cows daily consume about 4% of their body weight (~1,400 lbs/635 kg) daily—a whopping 56 lbs (24.5 kg) of grain every 24 hours.

(clker.com)

SI.2 Grain Consumption, Cattle

It is inefficient and nutritionally wasteful to feed grain to farmed animals[2]—and unethical given that 854 million people worldwide suffer from hunger on a regular basis ("Grain Harvest Sets Record"). Consuming animal products (rather than directly eating grains) wastes 80–90 percent of the protein that grains hold, 90–96 percent of grain's calories, and 100 percent of their carbohydrates and fiber (Kaufman and Braun 18). Furthermore, a steer/heifer does not merely grow, he or she moves about, plays with others (if possible), and continually burns calories simply to keep warm, digest, and grow and maintain hair, hooves, bones, and other "inedible" body parts (Figure 1.3). Consequently, most grains consumed by bovines do not convert to anything that omnivores place on their plates. Bovines eat about 20 pounds (9 kg) of food to gain just a couple of pounds in body weight—only a very small portion of which is gained in "eatable

---

2. While corn and soy are not complete food sources, neither is flesh. A balanced diet is essential to human health—but a balanced diet need not include *any* animal products. In fact, the leading health problems in the world's wealthiest countries (heart disease, cancers, stroke, respiratory disease—linked to diet through pollutions—and diabetes, not to mention obesity) are indisputably linked with a diet rich in animal products. Those who wish to live a long and healthy life, protect the environment, protect farmed animals, and/ or share with those who the hungry, will choose a plant-based diet.

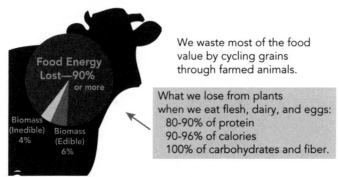

We waste most of the food value by cycling grains through farmed animals.

Food Energy Lost—90% or more

What we lose from plants when we eat flesh, dairy, and eggs:
80-90% of protein
90-96% of calories
100% of carbohydrates and fiber.

Biomass (Inedible) 4%

Biomass (Edible) 6%

Cattle, pigs, and poultry burn most grain calories in the process of living. Typically, less than 10% of what cattle consume becomes part of their physical body—much of which can't be eaten: Cattle consume ~ 5 metric tons to yield just 300 kg of edible flesh.

FIGURE I.3 What Is Lost?

flesh." At least 42 percent of a bovine's carcass is inedible ("How Many People"; "How Much Meat Will"). Using this figure, a steer of 1,250 pounds (567 kg) yields a *maximum* of 725 eatable pounds (329 kg), or more reliably, 600 eatable pounds (272 kg) ("How Much Meat Is"; "How Much Meat Will").

Why do ranchers and farmers choose to feed grain to bovines even though this option is extremely wasteful and even though we live in a world where thousands of people starve to death daily for want of grains? First, ranchers and farmers feed grain to farmed animals because consumers buy animal products. Second, we are currently raising many, many more farmed animals than on the earth's pasture lands can support. Grass fed beef is an unrealistic, selfish, elitist idea, accommodating only those who are comparatively rich—and unwilling to change their eating habits. Third, opting to feed grains yields a faster turnaround, which bolsters profits. For example, industry experts recommend that bovines be fed grain for 2–5 months before slaughter in order to speed the growth process and maximize profits ("Feedlot Industry"):

[C]alves that weigh less than 450 pounds [204 kg] perform poorly on grass pasture. If cattle weigh over 750 pounds [340 kg] in the spring, they should not be fed on pasture. It is recommended that they be placed in a confined situation and fed a higher proportion of concentrate feeds, such as barley or corn. (Comerford et al.)

Moving a calf from 500 pounds (227 kg) to 1,300 pounds (590 kg) in "just a few weeks"[3] requires grain (Comerford et al.). How much grain do calves consume before slaughter? Calves "will consume approximately 2.5 percent of their body weight each day (dry matter pounds)," which comes to about 20 pounds (9 kg) of feed per day for an 800 pound steer (363 kg) (Naze et al.; ZoBell et al., 4). As bovines grow, grain is increased "until the ration is approximately 65% concentrate [grain], 30% roughage and 5% supplement, fed 2 to 3 times per day" (Comerford et al.). This means that each feedlot calf is fed about 13 pounds (6 kg) of grain daily for at least two months, during which time one steer/heifer consumes a *minimum* of 780 pounds (354 kg) of grain. Most calves are confined in feedlots for four months, consuming more than a ton (907 kg) of grain.

Counterintuitively, fishmeal is also fed to farmed animals. Factory farms provide a dependable outlet for bycatch (described in the next chapter): "90 percent of the global small fish catch—which includes anchovies, sardines, and mackerel—is processed into fish meal and fish oil and used in animal feed" (Sachs; "A Broken Food Chain"). Worldwide, close to 8 million metric tons of fishmeal is used to fatten farmed animals— likewise an environmentally disastrous choice—as explained in the next chapter (Steinfeld et al. 12; Dawn 275)).

Consuming dairy products is also much more harmful to the environment than choosing a plant-based diet. Cows raised to produce milk require and consume more calories than do those raised for roasts or steaks because they must create and birth offspring in order to lactate. Some lactating cows consume more than 4 percent of their body weight, which is about 1,400 pounds for Holsteins, or a whopping 56 pounds (24.5 kg) of grain daily (Figure 1.4) (Grant and Kononoff). The food value and calories contained in this massive amount of grain—expended for just one lactating cow—cannot be retrieved by consuming cheese

---

3. Feeding grains moves "weaned calves (450 to 600 pounds [204–272 kg]) and yearling steers or heifers (550 to 800 pounds [249–363 kg]) to slaughter weights of 1,100 to 1,400 pounds" (499—635 kg) in just a few weeks during their final stage of growth (Comerford et al).

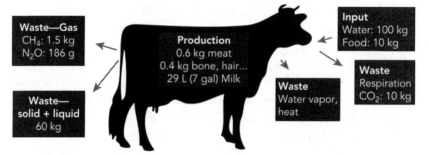

For every 100 kg of water and 10 kg of food consumed by a lactating cow, **80 kg are excreted or exhaled** (urine, feces, gas, water vapor). Because lactation is energy intensive, cows in dairy production eat more, drink more, and excrete more than other cattle.

("How Much Feed," McKague, Hawthorne, Kinsman, Voisinet).

FIGURE 1.4 What Goes in Must Come Out: Daily Dairy Input and Output

and yogurt—or by eating the hamburger she will eventually become. Furthermore, preparing the soil, planting, tending, and harvesting so many tons of grain to feed bovines wastes fossil fuels and emits a tremendous amount of GHGE.

Because feeding grains to farmed animals wastes more calories than are ultimately produced, only 37 percent of the current cultivated croplands would be required if we shifted to a plant-based diet. A professor of Ecology has described "the 10% rule" of trophic interactions this way:

> There is on average a 90% loss of energy as we go up each trophic level. Every 10 tonnes of grain produces just one tonne of meat (flesh). Looking only at our energy needs, 1 kg of grain can supply approximately what's in 1 kg of meat. So we only need 10 percent of lands currently cultivated for animal feed if we move to a plant-based diet. In principle it is that simple. We need the 30 percent of lands already cultivated for non-meat products (vegetables and grains) plus one tenth of the 70 percent currently used for animal agriculture. This leaves us with just 37 percent of the lands currently under cultivation if we quit animal agriculture. (Halley)

We would return more than 60 percent of cultivated lands to their natural state—or commit them to other uses—if this nation were to choose a plant-based diet. We would simultaneously markedly cut back on our use of fossil fuels, as well as our use of water, land, pesticides, chemical fertilizers, and so on.

Despite the hefty costs of fossil fuels, governments artificially lower petroleum prices, preventing industries and consumers from paying these high costs. (This government petroleum subsidy is particularly unjust given that disadvantaged people tend to live without automobiles. For example, in the U.S. some 7.5 million households in the largest metropolitan areas have no access to a private car, including almost 57 percent of households in New York City [Tomer; Schmitt].) Artificially lowering fuel prices benefits oil barons, other wealthy citizens, and those individuals and industries using more than their share of fossil fuels, including ranchers and Big Agriculture: "Intensive confinement operations require vast amounts of fossil fuel-based energy to cool, heat, and ventilate the facilities, and energy is also used to operate farm machinery to cultivate and harvest feed crops" ("Greenhouse Gas Emissions"). Instead of allowing the market to steer consumers and industries away from petroleum (and yogurt and bacon and omelets), government "subsidies strongly distort prices at the input and product level" for animal agriculture in "most developed and some developing countries" (Steinfeld et al. 232). Subsidies constitute a whopping "31 percent of farm income" globally (Steinfeld et al. 232).

Artificially lowered prices mask the actual cost of animal products, thereby fostering GHGE. Around the world, thanks to government subsidies, a diet rich in animal products is much more affordable than it would otherwise be—and more environmentally devastating. Worldwide, animal agriculture contributes more carbon dioxide to the atmosphere (through the use of fossil fuels) than any other single source (Goodland and Anhang 11). Animal agriculture "creates more global warming than all the cars, planes, trains, buses, and trucks in the world combined" (Oppenlander 16). Animal agriculture pumps *at least* 30 billion metric tons of carbon dioxide into the atmosphere each year, producing about 51 percent of our annual GHGE (Figure 1.5) (Goodland and Anhang 11). "The correlation is simple, direct, and irrefutable: Fewer animals raised for food means lower carbon emissions" (Cassuto 19). We can feed ourselves much more efficiently— and therefore with a much smaller environmental footprint—if we replace foods that contain animal products with fruits, vegetables, and grains.

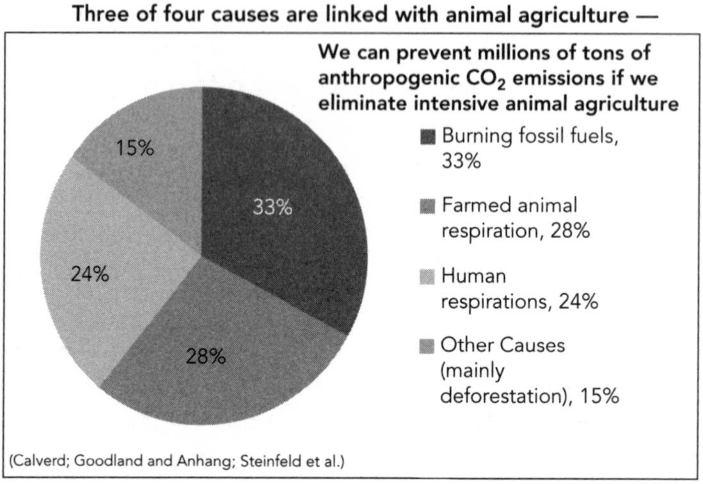

Three of four causes are linked with animal agriculture —

We can prevent millions of tons of anthropogenic CO_2 emissions if we eliminate intensive animal agriculture

■ Burning fossil fuels, 33%

■ Farmed animal respiration, 28%

■ Human respirations, 24%

■ Other Causes (mainly deforestation), 15%

15%

33%

24%

28%

(Calverd; Goodland and Anhang; Steinfeld et al.)

Release of CO_2 through respiration depends on energy use, which depends on biomass. Therefore, a species' CO_2 emissions can be estimated via species' biomass.

FIGURE 1.5 Key Anthropogenic Carbon Dioxide Sources Globally

Methane (CH_4)

Following the use of fossil fuels, methane (CH_4) is our second largest—and more potent—anthropogenic source of GHGE (Oppenlander 6). Methane remains in the atmosphere for nine to fifteen years, but in that time traps solar radiation 25 times more effectively than does carbon dioxide (Goodland and Anhang 13). This means that methane holds 72 times more heat than carbon dioxide when calculated across just twenty years (Steinfeld et al. 82; Oppenlander 6; "Methane vs. CO2"). Global warming, with the melting of permafrost and the ocean's icy crust, has released and continues to release methane previously sealed into the earth's surface or deep underwater. Around the world, methane pollution has increased 145 percent in the last fifteen years (Steinfeld et al. 114).

Animal agriculture is the largest human-induced (anthropogenic) source of methane, responsible for a whopping 40 percent of global emissions (Oppenlander 6; Steinfeld et al. 82, 95, 112). Both manure breakdown and enteric fermentation, a digestive process in cud-chewing animals such as bovines, sheep, and goats, release methane into the atmosphere. As a result of enteric fermentation, ruminants produce almost 8 million metric tons of methane annually—80 percent of agricultural methane production (Steinfeld et al. 112). Due to methane's potency, this provides the GHGE equivalent of more than 16 million metric tons of carbon dioxide (Cassuto 5; Steinfeld et al. 96). "Globally, ruminant livestock emit about 80 million metric tons of methane annually, accounting for 28 percent of global methane emissions from human-related activities" ("Enteric Fermentation Mitigation"). Bovines, both in the beef and dairy industries, account for 95 percent of these methane emissions ("Enteric Fermentation Mitigation"). The United Kingdom's 10 million cows produce 25–30 percent of Britain's methane pollution ("Future Technology"). A 10,000-cow dairy farm (larger dairy farms contain more than 15,000 cows) emits 33,092 pounds of methane every day (Hawthorne 37; "Changes" 2).

Those consuming grass-fed beef are responsible for even more methane emissions. As noted, it takes longer to fatten grass-fed calves for slaughter, and because grass is more difficult to digest, grass-fed bovines emit 50–60 percent more methane than grain-fed bovines (Oppenlander 125). In contrast, plants do not release methane (Figure 1.6) (Oppenlander 18). Animal agriculture puts methane, the most potent and dangerous GHGE, into the atmosphere both directly (via farmed animals) and indirectly (via carbon dioxide emissions that warm the planet and release methane from frozen lands and waters.

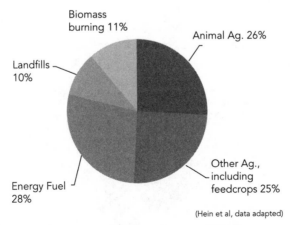

(Hein et al, data adapted)

FIGURE 1.6 Key Anthropogenic Methane Sources Globally

Nitrous Oxide (N₂O)

Nitrous oxide is the most potent anthropogenic ozone-depleting substance ("NOAA Study"). Nitrous oxide traps solar radiation 300 times
more effectively than does carbon dioxide, and stays in the atmosphere
"for an average of 120 years" ("Nitrous"). Worldwide, agriculture produces
a startling 65 percent of human-induced nitrous oxide (Oppenlander 6;
Steinfeld et al. 114).

Animal agriculture worldwide is responsible for 75 percent of these
emissions, resulting in the CO_2 equivalent of 2 billion metric tons of
GHGE ("Role of Livestock"). Nitrous oxide is produced when nitrogen is
added to the soil via synthetic fertilizers. In the U.S., "agricultural soil
management" (the use of synthetic fertilizers) is the largest source of
nitrous oxide emissions, accounting for almost 70 percent of total U.S.
N₂O emissions ("Nitrous Oxide Emissions"). Nitrous oxide is also produced through fossil fuel combustion and when manure decomposes (oxygen combines with nitrogen, emitting nitrous oxide). With a worldwide
increase in animal agriculture, the negative effects of manure decomposi-
tion have increased proportionally (Steinfeld et al. 114). Animal agriculture
creats 96 percent of the earth's nitrous oxide (Figure 1.7) ("Nitrous Oxide
Emissions"; Gluckman; "What Are"). A modest 10,000-cow dairy farm
emits 409 pounds of nitrous oxide every day (along with 3,575 pounds of
ammonia) (Hawthorne 37; "Changes" 2). All told, U.S. animal agriculture produces almost 1.4 million metric tons of nitrous oxide every year,
providing the GHGE equivalent of more than 37 million metric tons of
carbon dioxide (Cassuto 5).

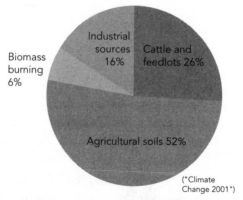

FIGURE 1.7 Key Anthropogenic Nitrous Oxide Sources Globally

Nitrous oxide also combines with oxygen to create nitric oxide—which converts to nitric acid, falling to the earth as acid rain, damaging and destroying landscapes ("Acid Rain"). (More information is provided on acid rain in this chapter under the heading, "Water Pollution.") Consuming plant-based foods, which greatly reduces fossil fuel combustion, manure production, and the need for monocultures for feedstock, minimizes anthropogenic nitrous oxide emissions (Oppenlander 18). Consuming animal products greatly increases already high anthropogenic carbon dioxide, methane, and nitrous oxide emissions (Figure 1.8). Switching to a diet of grains and greens is the best way to reduce our GHGE (Figure 1.9).

Water Pollution

Agribusinesses causes "more water pollution than all other industrial and municipal" sources of water pollution combined (Cassuto 9). Animal agriculture creates twice as much freshwater pollution as do other industries, and five times as much as do households (Schwartz 87). "Big Ag accounts for 48 percent of the pollution found in streams and rivers and 41 percent of pollution in lakes" (Hawthorne 36). Thanks to animal agriculture, our water contains an abundance of manure, pesticides, antibiotics, nitrates, and arsenic, much of which runs into the seas.

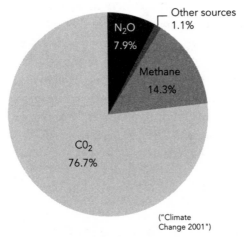

("Climate Change 2001")

FIGURE 1.8 Anthropogenic Greenhouse Gas Emissions Worldwide

	World Wide emissions (Metric tons)	GHGE CO$_2$ Equivalent (Metric tons)
Carbon Dioxide	8-15 billion	8-15 billion
Methane	68 million	1.5 billion
Nitrous Oxide	8.2 million	2.4 billion

Source: EPA Energy Calculator (http://www.epa.gov/cleanenergy/energy-resources/calculator.html#results)

FIGURE 1.9 GHGE from Animal Agriculture Worldwide

Manure

Animal agriculture in industrial nations generates large herds of large animals who are trapped in small spaces. This unnatural situation creates waste disposal problems. U.S. animal agriculture produces 454 million metric tons of animal waste every year—some 1.3 billion metric tons of wet waste (Schwartz 87) and 304 million metric tons of dry waste (sans urine and fluids)—contributing "almost a third of the total municipal and industrial waste produced every year" (Figure 1.10) (Gluckman; "Final Report"). Farmed animals in the U.S. create 130 times the amount of human waste processed in U.S. municipal wastewater plants each year . . . but agricultural waste is not cycled through a treatment plant (Oppenlander 54; "Final Report").

Each cow on a dairy farm releases almost 150 pounds (17 or 18 gallons) (68 kilos/67 liters) of waste every day (Hawthorne 41; Chastain and Camberato).[4] A dairy with a mere 2,500 cows produces roughly as much waste as a city of 411,000 people ("Final Report" 20). Pigs are the biggest poopers, producing 80,000 pounds (32,300 kg) of waste every year per 1,000 pounds (454 kg) of pig (Gaechter). A farm with 5,000 pigs produces as much raw sewage as a town of 20,000 humans (Cassuto 7). There are more than 60 million pigs in the U.S. Smithfield Farms, the world's largest pork producer, "discharges" almost 24 million metric tons of waste every year (Dawn 276–77). "Discharges"— where do you suppose it goes?

---

4. 12 gallons (45 liters) of waste per 1,000-pound (454 kg) bovine per day, with Holstein cows weighing about 1,400 pounds (Chastain and Camberato).

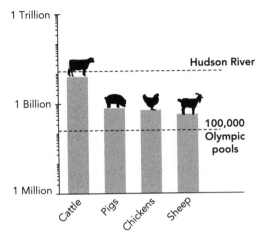

FIGURE 1.10 Solid Waste of Domestic Animals (tons/year)

Animal agriculture facilities often store manure in open lagoons, some of which inevitably spill over, leak, or burst, damaging and destroying ecosystems. News reports generally only mention these environmental catastrophes when water systems are directly affected. Here are some recent reports of manure spills:

5,000 pigs produce as much poop as do 20,000 humans.

In the U.S. > 60 million pigs produce 80,000lbs (32,300 kg) of slurry every year.

Being smaller, chickens produce less poop per individual, but each of nearly 16 billion chickens produces, on average, 30 lbs of poop annually. **The global flock releases about 480 billion lbs (218 billion kg) of waste every year** (Ritz and Merka "Chapter 6").

S1.3 Pig Poop

- Nine million gallons of chicken droppings escaped into North Carolina's Limestone Creek when a lagoon on a "laying hen" farm failed (Hawthorne 37).
- A 21-million-gallon (80-million-liter) manure lagoon on a dairy farm in Washington released "an unknown amount of manure into fields . . . and [into] the Snohomish River" ("Big Manure Spill").
- Nearly 30,000 fish were killed in Iowa when hog manure leaked into the South Skunk River ("Separate Hamilton County").
- A dairy farm in Virginia spilled at least 40,000 gallons (151,000 liters) of manure into the New River, killing pretty much everything in its wake (Heineck; Dawn 275).

One news report noted that 1.5 miles (2.4 km) of roadway had to be closed in Maryland while crews cleaned up manure from an "unidentified source" ("Manure Spill Over"). (Other than animal agriculture, is there *any* source for such massive quantities of manure? Why such hesitance to state the obvious?)

Water pollution from manure is exacerbated by the use of fertilizers. More manure is produced by animal agriculture than farmers can use, so they spread excessive quantities of manure on fields as fertilizer, where they are carried into the local water systems by rainfall. For example, when heavy rains carried manure from a hog lagoon into Indiana's Mississinewa River, the problem was exacerbated by 27,000 gallons (102,000 liters) of manure that had been applied to a nearby field, which also washed into the water system. Some 40,000 fish were killed, along with other local wildlife (Richmond).

Manure is an environmental hazard even if contained. As manure decomposes, it creates dangerous quantities of nitrogen, ammonia, arsenic, and nitrate.

- Nitrogen: Decomposing animal waste emits nitrogen, which becomes nitrous oxide. Nitrous oxide is not only a potent GHGE; it also combines with oxygen to create nitric oxide—which converts to nitric acid, falling to the earth as acid rain. Acid rain damages and destroys soils, lakes, and forests, which harms and kills wildlife ("Acid Rain"). Forty-one percent of the Adirondack Mountain lakes in New York "exhibit signs of chronic and/or episodic acidification" ("Nitrous Oxide Emissions"; Driscoll et al.).

- Ammonia: Some 94 percent of human induced ammonia emissions worldwide stem from agriculture, 68 percent of which stem directly from farmed animal manure (Driscoll et al.; Steinfeld et al. 114). Manure emits "a total of 30 million tonnes of ammonia" annually (Steinfeld et al. 272). Ammonia damages plants and causes dead zones.
- Arsenic: Chickens are fed arsenic (to kill parasites and promote growth), which is then emitted in their waste, raising arsenic levels in local water and soils. Arsenic is poisonous, significantly reducing biodiversity in the region, while causing cancers and gangrene in human beings ("Arsenic").
- Nitrate: Chicken waste contains high levels of nitrate. Near chicken farms, water sources are often polluted with nitrate to the point where water is dangerous for animals, including human beings.

Animal waste pollutes water, destroying wildlife, damaging ecosystems, and harming human beings. Pollutants from farmed animal manure have been linked to irreversible brain damage (hydrogen sulfide poisoning), miscarriages, acute gastroenteritis, kidney failure, and infant deaths (from nitrate poisoning, i.e. methemoglobinemia or "blue baby syndrome") (Hawthorne 36, 41). Despite this, many factory farms (and gigantic open-air sewage ponds) stand just 75 feet (23 meters) above vital freshwater sources such as the Ogallala Aquifer. Despicably, factory farms are also frequently located in disadvantaged communities, where people tend to be less able to defend their health interests against the forces of Big Ag (Kemmerer 15).

Dead Zones

In a process of human-caused "eutrophication," manure and chemical fertilizers are flushed into waterways, where they create algae blooms. As these algae blooms decay, they consume oxygen, which creates conditions of anoxia (oxygen depletion). While this process is natural over long periods of time, on a limited scale, in lakes, rivers, and other small bodies of water, accelerated eutrophication caused by the growth of animal agriculture pollution in the last fifty years has brought about an entirely new phenomenon—dead zones. Massive pollution from industrial agriculture has created super-eutrophic zones that annihilate life in large lakes and throughout vast stretches of ocean, called dead zones, that cannot sustain life.

Dead zones were first recorded as far back as the early 1970s, when they appeared every two or three years. The UN recently identified a whopping 146 dead zones (and rising), most of which appear annually—in conjunction with agricultural runoff. Other researchers now track more than 400 dead zones around the world (Figure 1.11) (Perlman). Dead zones are visible in the Chesapeake Bay and along 300 square miles (777 km²) of the Oregon coast. The Baltic Sea suffers the largest dead zone—27,027 square miles (70,000 km²). Thanks to agricultural drainage in the Mississippi River, the Gulf of Mexico supports the second largest dead zone, affecting roughly 8,000 square miles (20,700 km²) (Steinfeld et al. 212). More recently, dead zones have emerged in South America, China, Japan, southeastern Australia, and the comparatively pristine little nation of New Zealand, known for animal agriculture. Imagine driving through vast areas, recently stripped of life. Could we but see what animal agriculture causes under the ocean's shining surface, citizens would surely respond—there would be an outcry. But when we turn toward the sea, our vision only skims the surface. Except when sea life washes ashore, dead zones remain hidden from view.

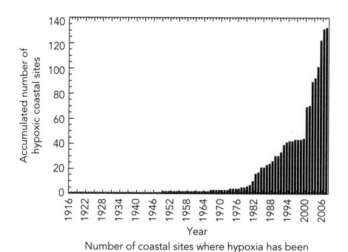

Number of coastal sites where hypoxia has been reported, 1916–2006.

Exponential growth rate = 5.54% ± 0.23% year−1
(R2 = 0.86, P ≤ 0.01)

Ref: Vaquer-Sunyer and Duarte, PNAS, 2008; 105:15452-15457

FIGURE 1.11 Increase in Coastal Dead Zones, 1916–2006

Other Pollutants

Manure and urine contain a flush of additional pollutants that find their way into water systems, including antibiotics, pesticides, herbicides, hormones, and cleaning solvents. These enter water systems, move through food chains, and become concentrated in those who eat flesh—predators, scavengers, and omnivores (including human omnivores) (Hawthorne 36; Dawn 274). An estimated 70 percent of U.S. antibiotics—about 29 million pounds (13 kg) every year—are fed to farmed animals . . . and wind up in the water (Hawthorne 38). Dairy cows are fed hormones to increase milk production; in feedlots, bovines are fed growth hormones. Much of what farmed animals consume is carried out in waste and into the water system. In water downstream of a beef feedlot where growth-promoting hormones were fed to cattle, female fish were found to have physical characteristics of breeding males, while males had reduced sex characteristics as a result of excessive male hormones in the water ("Hormones Used").

Monocultures, grown largely to feed farmed animals, require more pesticides, herbicides, and chemical fertilizers than do traditional farming methods, so it is not surprising that pollutants have skyrocketed with the growth of Big Ag; pesticide use has increased 400 percent in the last fifty years (Halden and Schwab 8; Kaufman and Braun 20).

Body parts such as blood, fat, and hair also flow into our water system from animal operations, especially slaughterhouses. Attempting to curb this problem, industry experts recommend methods for screening out at least the big chunks when flushing slaughterhouses. They also note possible problems and likely solutions:

1. The blood from slaughtered animals will coagulate into a solid mass, which may block up both open and closed drains. It is therefore recommended that the blood is collected and used for human consumption, stockfeed production or fertilizers, if the religious and cultural traditions allow the use of blood.
2. Solids (meat or skin trimmings, hair, pieces of bones, hooves, etc.) must be screened . . . by providing the drains with vertical sieves.
3. Effluents from slaughterhouses always contain small amounts of fat (melted fat or small pieces of fatty tissues). Grease traps should be installed in the drains [where] fat solidifies, rises to the surface and can be removed regularly. ("Hygiene")

Despite these suggestions, employees know that the "easiest disposal method is to divert effluents into existing pools, rivers or lakes" ("Hygiene").

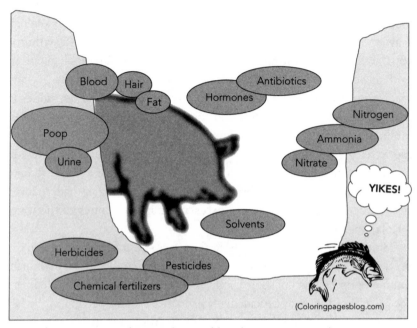

SI.4 What Does Animal Agriculture Add to the Water System?

Can I offer you a tall, cool glass of water?

Freshwater Depletion

Critical sources of freshwater are drying up around the world. Life-sustaining waterways such as the Colorado, Nile, and Yellow Rivers now run dry before they reach the sea (Brown 75–76). Animal agriculture is a key factor in this depletion of freshwater. For example, on behalf of corporate agriculture, the U.S. House of Representatives recently passed a bill "siphoning water from the Sacramento-San Joaquin Delta, where water levels are already dangerously low" (Pepino 6). The Colorado, recently named "America's most endangered river," is also suffering from "skewed allocation rights and water diversions" (Pepino 6).

Important underwater aquifers such as the Ogallala Aquifer are also shrinking. The Ogallala Aquifer spans "800 miles [1,287 km] from north to south, and 400 miles [644 km] from east to west," supplies 30 percent of U.S. groundwater irrigation to 27 percent of the nation's irrigated land, and is critical to an area often described as "America's Breadbasket"— though this area would more correctly be termed "America's Fleshpot." The Ogallala provides irrigation for 27 percent of the Nation's watered agricultural lands, and drinking water for 82 percent of the people living in South Dakota, Nebraska, Colorado, Wyoming, Oklahoma, Texas, New Mexico, and Kansas (Litke; Worm). But the Ogallala Aquifer is more than half gone, and looks likely to run dry in the next twenty five years.

Animal agriculture drains freshwater in several key ways. First, farmed animals require freshwater. And thanks to consumer choices, the earth currently supports at least

- 16 billion chickens,
- 2 billion pigs, and
- 1.4 billion bovines ("Counting Chickens," "Pig Facts and Pig Trivia").

Average water consumption per individual per day is:

- chicken, 1.5 cups (0.35 liters);
- hog, 3 gallons (11.4 liters);
- bovine, 37 gallons (140 liters).

This indicates that each day, according to species, the global herd's water average consumption is roughly as follows:

- chickens, 1.52 billion gallons (5.75 billion liters);
- hogs, 6 billion gallons (22.7 billion liters); and
- bovines, 52 billion gallons (197 billion liters).

Annual average water consumption per species rounds out to:

- chickens, 550 billion gallons (2082 billion liters);
- hogs, 2190 billion gallons (8289 billion liters); and
- bovines, 18,907 billion gallons (71,563 billion liters).

Combined, on average, chickens, pigs, and bovines suck up at least 21,647 billion gallons (81,943 liters) of freshwater every year. To put this in perspective, an Olympic sized swimming pool holds 660,430

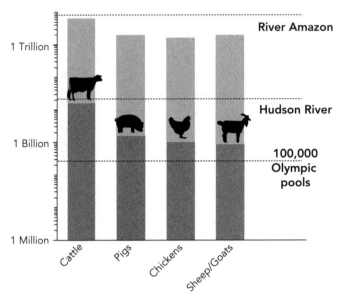

FIGURE I.I2 Water Consumption by Domestic Animals (tons/year)

gallons (2,500,000 liters) of water, and there are *more than* 1,500 Olympic swimming pools in just one billion gallons of water. Every year the world's bovines, pigs, and chickens, drink an average of about 33 million Olympic swimming pools of freshwater, and this does not even count the many goats, turkeys, and sheep who also turn to troughs for refreshment (Figure 1.12).

Animal agriculture's most excessive squandering of freshwater is not the global herd's water requirement but their food intake—irrigation for feed crops. Not only do we feed 70 percent of U.S. grains to farmed animals, but we waste "1000 tons of water to produce one ton of grain" (one U.S. ton = 907 kg) (Brown 71). This extravagance is aggravated by the fact that grains are often grown in arable lands (Steinfeld et al. 134) (Figure 1.13). Agriculture is responsible for 70 percent of the world's freshwater expenditure, and "93 percent of water depletion worldwide" (Steinfeld et al. 126). Slaughterhouses and dairies also require large quantities of freshwater. These facilities are cleaned several times each day—literally blasted with water, while "[w]ater tables are falling on every continent" (Brown 75–6). Dependence on animal agriculture causes "severe environmental degradation through water depletion" (Steinfeld et al. 134).

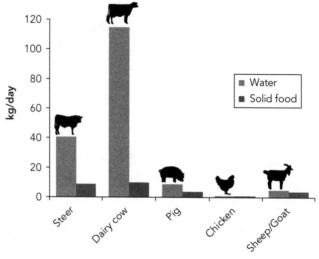

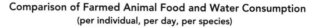

FIGURE I.13 The Big Gulp

Because of the amount of freshwater required for animal agriculture, the average American omnivore "consumes" 4,200 gallons (15,900 liters) of freshwater per person per day, while an American vegan averages just 300 gallons (1,136 liters) per person, daily (Schwartz 86). Consider these comparisons:

- 5,200 gallons (19,700 liters) of water are necessary to produce one pound of beef, but only 23 gallons of water are necessary to produce one pound of tomatoes (Cassuto 9).
- 500–2,500 gallons (1,893 to 9,464 liters) of water are necessary to produce one pound (0.45 kg) of beef, but only one-hundredth as much water is necessary to produce a pound of wheat (Kaufman and Braun 21).
- 5,200 gallons (19,700 liters) of water are necessary to produce one pound of California beef, but only 25 gallons of water are necessary to produce one pound of California veggies (Schwartz 86).
- Roughly 100 times more water is necessary to produce one pound of beef than is necessary to produce one pound of potatoes (Schwartz 86).

Why bother to shorten your shower if you plan poultry or pot roast for supper?

Land: Deforestation, Soil Degradation, and Land Use

Animal agriculture harms forests, soils, and public lands. As with GHGE and water depletion, animal agriculture is the number one culprit with regard to deforestation and soil degradation. In the U.S., cattle ranchers in particular usurp and damage public lands in their quest for private profits.

Deforestation

For the sake of grazing and raising feed for farmed animals—for the sake of such things as ham, cheese spreads, steak, egg sandwiches, and milk-shakes—one fifth of the world's rainforests were destroyed between 1960 and 1990. Between 1985 and 1990, 210 million acres of forest were turned to pasture, "an area nearly the size of Texas and Oklahoma" (Kaufman and Braun 18). Still today, a section of rainforest about the size of 20 football fields (22 soccer/football fields) is destroyed roughly every minute of every day, and in "the Amazon, cattle ranching is now the primary reason for deforestation" (Steinfeld et al. 272). In just fifty years, 50 percent of Costa Rica's forests disappeared—60 percent were cleared for bovines

Worldwide:
The primary cause of deforestation is conversion of lands for feedcrops and grazing.

Brazil is one of the world's leading
- beef producers,
- nations suffering from deforestation,
- producer of soybeans for feedcrops.

55 feet (17 m) of tropical forest yields a mere ¼ lb (120 g) hamburger.

(wpclipart.com)

SI.5 Deforestation Details

("Deforestation in Costa Rica"). Only 13 percent of Costa Rica's original rainforest remains, and what remains is now in a "highly fragmented and degraded state" (Reynolds and Nierenberg 11).

Brazil continues to lead the way (by a considerable margin) in rainforest destruction; agriculture is responsible for roughly 98 percent of Brazil's deforestation—ranchers are responsible for 65–70 percent of Brazil's lost forests (Butler). There were about 10 million bovines in Brazil in 1980, now upwards of 55 million ("Deforestation: The Leading Cause"). South America is still at the top of the list for loss of forests, but the U.S. and the EU are implicated: The U.S. imports some 80 million pounds of Brazilian beef every year; 85 percent of EU beef originates in Brazil. All this ecological devastation for a mere spot of flesh—55 square feet (17 square meters) of tropical forests yield just a quarter pound (120 grams) of hamburger. If we continue as we are, primary forests will be altogether gone by 2050 (Hawthorne 39; Pimm and Raven 844).

Not only beef eaters but those who consume turkeys, pigs, chickens, eggs, and/or dairy products are implicated. The primary reason for loss of forests is conversion of land to agriculture—both for grazing *and for feed crops*: In Latin America, land is converted from forests to agriculture largely for feed crops, "notably soybeans and maize" (Steinfeld et al. 12). Brazil's soy crop grew more than 3,000 percent in the last forty years, making the country the world's second largest soybean producer ("Agriculture in Brazil"). Worldwide, 80 percent of the soybean crop is fed to farmed animals (Reynolds and Nierenberg 13); in the U.S. 98 percent of soy is turned into meal to feed poultry, bovines, hogs, and farmed catfish, while more than 50 percent of U.S. soy produced is fed to poultry. Those who cast an accusing eye at soy-eaters have missed a vital point: *Worldwide, 80 percent of soybean crops are planted, tended, and harvested for farmed animals, implicating those who eat cheese and chickens—not those who eat tofu and tempeh* (Reynolds and Nierenberg 13).

For purely selfish reasons, most of us do not wish to see rainforests disappear. Rainforests hold a good deal more than half of the world's animal and plant species—if we lose most of the rainforests, we lose at least 50 percent of earth's species (Pimm and Raven 844; "Rainforest Facts"). Many of these species have not even been classified by Western scientists—plants and animals rich with medical and nutritional possibilities remain safely hidden from chemists and cooks. Rainforests harbor untapped recreation sites for flush travelers, and protect some of the most hidden and "untainted" human cultures. Additionally, rainforest soils are critical to land stability—devoid of rich canopy and reaching tree roots, soils are prone to mudslides and desertification. Also of critical importance, rainforest tree canopies are powerful converters on behalf of the atmosphere, turning carbon dioxide into oxygen, helping to mitigate some of the effects of climate change.

Forests are home to 70 percent of the earth's land animals and plants. Estimates of the rate of species extinction vary widely, but conservative estimates indicate that we are losing at least 137 species every day ("Rainforest Facts"). Forests are also important to the earth's water systems, holding water and generating rain for drought-plagued landscapes ("Forest Holocaust").

- Hold half the world's species
- Protect hidden indigenous peoples
- Foster innumerable medical possibilities
- Mitigate climate change
  ...and are precious in their own right

(wpclipartcom)

si.6 Rainforests

According to the World Resources Institute, more than 80 percent of the Earth's natural forests already have been destroyed. Up to 90 percent of West Africa's coastal rain forests have disappeared since 1900. Brazil and Indonesia, which contain the world's two largest surviving regions of rain forest, are being stripped at an alarming rate by logging, fires, and land-clearing for agriculture and cattle-grazing ("Forest Holo caust").

The "leading cause of this deforestation is meat production" (Hawthorne 39). We are chewing anteaters, armadillos, jaguars—and plants and animals we have not yet noticed—into oblivion.

Soils

Worldwide, the leading causes of soil degradation are overgrazing (35 percent), deforestation (30 percent), and large-scale agriculture more generally (28 percent)—all directly linked to our consumption of animal products (Figure 1.14) ("Land Degradation"). Forty percent of the world's agricultural lands were degraded in the last century "by the hard hooves and heavy bodies" of millions of farmed animals (Wardle). Dry areas have suffered the most—73 percent of dry rangeland worldwide is already degraded (Steinfeld et al. 272). Farmed animals are the primary cause of desertification, both through overgrazing and from the production of feed crops—especially monocultures ("Public Lands Ranching").

Many lands are not suited for grazing, and therefore can only be turned to grazing for a handful of years before they are depleted of nutrients, at which point they become wastelands (Hawthorne 39). When grazing lands become wastelands, people clear and till neighboring plots of land in order to continue production. But these lands, too, can only sustain grazers for a short period of time. Consequently, "every year a new chunk of real estate the size of Rhode Island is being swallowed by sand" (Hawthorne 40). Depleted soils from these dusty graveyards blow across continents. Seoul, for example, often falls under a perpetual brown haze caused by China's newly created northern wastelands—the result of overgrazing in the hope of satisfying China's growing demand for flesh.

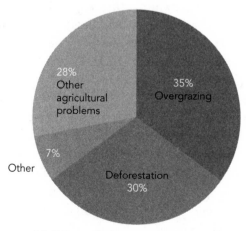

FIGURE 1.14 Causes of Soil Degradation

Land Use

"The livestock sector is the world's largest consumer of the Earth's land resources" (Reynolds and Nierenberg 13). Worldwide, feed-crop production claims about 1,164 million acres (471 million hectares)—"33 percent of the total arable land" (Steinfeld et al. 272). About 8.6 billion acres (3.5 billion hectares) are devoted to grazing lands ("Ecological Footprint"). The amount of land tilled and tended on behalf of animal agriculture worldwide (both grazing and crops) is roughly the size of Africa (Hawthorne 39).

Animal agriculture requires monumentally more land than any other human enterprise (Steinfeld et al. 133). In the U.S., 84 million acres (34 million hectares) are devoted to growing corn, 80 percent of which is fed to farmed animals at home and abroad. An additional 74 million acres (30 million hectares) are devoted to soybeans. Another 53 million U.S. acres (22 million hectares) are put into wheat, some 22 percent of which is used for animal feed; 8 million acres (3.2 million hectares) are devoted to sorghum, almost all of which is fed to farmed animals, and 60 million acres (24 million hectares) are devoted to hay and alfalfa—all for farmed animals ("Major Crops"). This means that 58 million U.S. acres (23.4 million hectares) of corn, more than 22 million acres (9 million hectares) of soybeans, some 12 million acres (5 million hectares) of wheat, and about 7 million acres (2.8 million hectares) of sorghum are planted, irrigated, doused with pesticides, herbicides, and chemical fertilizers, harvested, transported, packaged, and stored in order to feed animals—just shy of 100 million acres (40 million hectares) are dedicated to crops raised not for human consumption but to be cycled inefficiently through farmed animals. U.S. animal agriculture claims an additional 525 million acres (212 million hectares) of land for grazing (Figure 1.15) ("Land Use").

Some 98 million bovines live in the U.S. When planted with soy and corn for human consumption, 2.5 acres (1 hectare) of land can produce 2,200 pounds (1,000 kg) of protein, but an unnerving 25 acres (10 hectares) are required to produce the same 2,200 pounds of protein from steers and cows. As noted, we would need only 37 percent of current croplands if we were not feeding grains to farmed animals. A plant-based diet "uses less than half as many hectares as grass-fed dairy and one-tenth as many hectares as grass-fed beef to deliver the same amount of protein" (Matheny). If we adopt a plant-based diet, hundreds of millions of acres can be returned to wildlands—forests and grasslands and prairies—an environmentalist's dream come true.

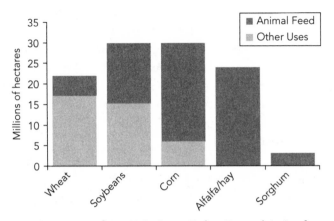

100 million acres (40 mil hect) of U.S. land are Devoted to feedcrops

FIGURE 1.15 Percentages of Key U.S. Crops Fed to Farmed Animals

Wildlife

What is the common denominator in the disappearance of Wyoming's bald eagles, black-footed ferrets, swift fox, long-billed curlews, ferruginous hawks, burrowing owls, upland sandpipers, white- and black-tailed prairie dogs, lynx, Lewis's woodpeckers, black-backed woodpeckers, three-toed woodpeckers, and wolverines? Each of these species (and many others) are disappearing because Wyoming prairies (as well as grasslands in other states) are being converted to croplands and pasture for farmed animals, largely bovines, and/or because of "predator control" on behalf of ranchers (and hunters—see chapter 3 for more on predator control) (Nicholoff 16–28).

Aldo Leopold wrote, "[p]redatory animals are the common enemy of both the stockman and the conservationist" (Leopold, "Wanted" 451). On behalf of ranchers (and hunters) the government's Wildlife Services Predator Control Program has been wiping out predators since the early twentieth century. The agriculture lobby, which views large predators as a "threat to livestock," asks government wildlife agencies to eliminate any animals deemed hazardous to their investments—and the U.S. government's Wildlife Services willingly complies (Fox): Every year since 2004 somewhere between 100,000 and 125,000 mammalian carnivores have been killed. Every year at least since 2004, some 600 badgers, 400–500 black bears, 10,000–13,000 raccoons, and 70,000–90,000 coyotes have been killed by the U.S. government—at taxpayer expense—on behalf of agriculture interests ("Wildlife Killed"). Though "motivated to undercount the number of animals" they kill, in 2011 Wildlife Services predator control reported killing 11,061 birds—hawks, falcons, owls, and vultures—(up 42 percent from 2010) and 116,093 land predators (up 3 percent), including wolves, coyotes (83,695), bears, bobcats, fishers, cougars, weasels, skunks, raccoons, and foxes. Also in 2011, Wildlife Services poisoned 18,587 animals (up 31 percent, including an additional 2,300 coyotes), and no doubt at considerable expense, gunned down 48,811 animals from helicopters (a 15 percent increase from 2010) ("Wildlife Services"). The most recent report indicates that the U.S. federal government killed a whopping 3.8 million animals in 2011 (a 25 percent decrease from 2010 only because they killed roughly a million fewer starlings).

Traps and poisons are indiscriminate. Every year predator control kills thousands of animals incidentally, including armadillos, bluebirds, bears, bobcats, turtles, and alligators ("Table G-2"). In 2010, Wildlife Services' indiscriminate methods killed nearly 200 owls, nearly 400 falcons, and more than 1,000 hawks. Inevitably, among these casualties were threatened and endangered species, including nearly 2,000 grey wolves "unintentionally" killed by predator control since 2004 ("Wildlife Killed"). It is simply not possible to maintain the integrity of ecosystems while killing thousands of wild animals from specific species, yet U.S. Federal Wildlife Predator Control Program continues its war on wildlife—"at the request of ranchers," at the expense of taxpayers (Lange 18). And just for the record, predator control has also killed "nearly 1,400 house cats, [and] more than 400 domestic dogs" ("Wildlife Services").

To make all this killing possible, 53 percent of the Wildlife Services budget is spent "protecting livestock"; the U.S. government maintains a tax-funded "livestock protection budget" of more than $10 million (O'Toole). Adding insult to injury, "75 percent of federal livestock funds are spent on public lands" (Figure 1.16), including 40 percent of Wildlife Services' federal funds, which are "dedicated to the 27,000 ranchers who graze livestock on public lands" (see next section on public lands for more on this topic) (O'Toole). Wildlife Services handouts encourage ranchers to rely on federally financed killing to protect herds, rather than rethink land use and/or find ways to coexist with wildlife (O'Toole).

Yet more disturbing, all of this killing and populations manipulation is to no end (except, perhaps, for the handful of people who earn a living via

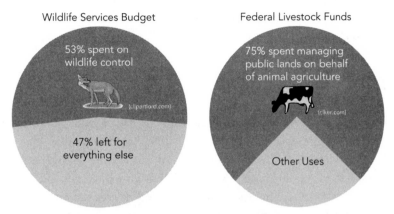

FIGURE 1.16 Public Cost of Animal Agriculture in the U.S.

"predator control"). Predator control is a waste of taxpayer monies because it offers only a very short-term fix. Killing members of target species usually causes a population to *increase* in short order, because births adjust for deaths. For example, even when more than half of a coyote pack is wiped out, the pack is likely to bounce back within a year (Lange 17). Though predator control is a very short fix (and counterproductive), and though this program damages ecosystems and further threatens endangered wildlife, the U.S. government continues its multimillion-dollar extermination program, decade after decade, on behalf of animal agriculture.

When ranching interests conflict with ecosystems and wildlife, "it is the industry, not the predators, that ought to go," but there are few industries as powerful as animal agriculture, especially ranchers (Regan 359). And in the U.S. "[b]eef production is the largest single segment of American agriculture. The U.S. beef industry is comprised of more than 1,000,000 businesses, farms and ranches with over 800,000 ranchers and cattle producers" ("Beef Cattle Production"). As a result, the economic interests of Big Ag trump the protection of ecosystems and wildlife, even threatening the Endangered Species Act "and all the creatures it protects" (Turner 11). In 2011, for example, Montana's governor, Brian Schweitzer, openly "ordered his wildlife officials not to enforce certain federal prohibitions on wolf killing in his state, and to kill entire wolf packs in response to even first-time wolf depredation on livestock. . . . And it isn't just wolves" that he considers expendable on behalf of Big Ag (Honnold and Harbine; "Howling"; Turner 10–11). "Keystone predators like the grizzly and Mexican gray wolf were driven extinct in southwestern ecosystems by 'predator control' programs designed to protect the livestock industry" ("Grazing").

U.S. Public Lands

Ranching "is the most widespread commercial use of public lands in the United States" ("Public Lands Ranching")—an environmental (and economic) catastrophe. With regard to public lands, Wildlife Services "creates perverse incentives for ranchers to use sub-marginal land, [and also] to overgraze" (O'Toole). Bovines harm native wildlife, vegetation, and soils. They do not browse and wander like native mammals, such as elk and deer. Bovines tend to "remain in the same area until they have eaten all or most of the edible material," trashing waterways, soils, and delicate vegetation in the process, and "causing significant harm" to native species and their ecosystems ("Cattle Grazing"; "Grazing").

Bovines are especially hard on wetlands and water systems. For example, they add manure and bacteria to waterfowl nesting habitat (increasing water temperatures), trample waterfowl nesting sites, and destroy water retention ("Cattle Grazing"). Bovines strip riparian (streamside) "habitat favored by many small mammals and nesting songbirds," destroying root systems that stabilize stream banks, leading to erosion and loss of pebble beds where fish spawn (Chadwick 26). Bovines

> destroy native vegetation, damage soils and stream banks, and contaminate waterways with fecal waste. After decades of livestock grazing, once-lush streams and riparian forests have been reduced to flat, dry wastelands; once-rich topsoil has been turned to dust, causing soil erosion, stream sedimentation and wholesale elimination of some aquatic habitats. ("Grazing")

Ranching is also a primary cause "of native species endangerment in the American West" ("Grazing"). Across time, innumerable plant species have been eliminated through overgrazing. In the arid Southwest, farmed animal grazing "is the most widespread cause of species endangerment" ("Public Lands Ranching"). Despite this plethora of serious and well-documented environmental problems, no "report has ever fully analyzed the incredible environmental costs of livestock grazing on federal public lands" ("Fiscal Costs").

The U.S. government's grazing program is also a financial disaster, losing "money just as rapidly and consistently as it destroys habitat" ("Grazing"). Permitting ranchers to graze farmed animals on federal lands (Bureau of Land Management [BLM], Forest Service) costs taxpayers as much as $1 billion annually. Indirect costs are likely about three times this amount, while ranchers enjoy "$100 million annually in direct subsidy" ("Grazing"). Administering public lands to benefit ranchers creates a deficit: The U.S. government spends at least $144 million managing farmed animals on federal lands—while collecting only $21 million in grazing fees—a net loss of $123 million (Figure 1.17) ("Fiscal Costs"). Grazing fees are so low on public lands—currently $1.35 for one month for one cow and her calf across 16 Western states (on public lands administered by the Bureau of Land Management and the Forest Service ["BLM and Forest Service"])—"that they amount to a subsidy" ("Cattle Grazing").[5] In contrast, "[p]rivate, unirrigated rangeland in the West rents out for an average of $11.90" per cow and calf—the federal grazing fee is "a *de facto* subsidy for cattle owners". . . along with artificially low fuel costs ("Grazing").

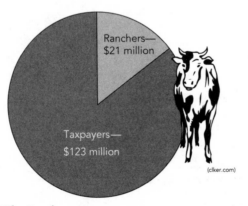

The US government spends > $144 million every year to manage farmed animals on federal lands.

Ranchers— $21 million

Taxpayers— $123 million

(clker.com)

FIGURE 1.17 Who Pays?

5. The BLM manages 245 million acres (99 million hectares) largely in 12 Western states— more than any other Federal agency—while the Forest Service manages 193 million acres (78 million hectares) in 44 states, Puerto Rico, and the Virgin Islands ("BLM and Forest Service").

On average, "three-fourths of all federal livestock funds, and three-fifths of cooperative funds, are spent on public lands" (O'Toole). As much as 96 percent of these government expenditures—which ought to enhance public lands—"enhance livestock production *in direct conflict with legal mandates to restore the health of public lands*" ("Public Lands Ranching," emphasis added). For example, in 2004 and 2005 in Arizona's Tonto National Forest, ranchers were given $3.5 million federal dollars for "range improvement" ("Grazing"); "range improvements" alter landscapes, damaging ecosystems, and are therefore not in the best interest of the tax-paying public.

The costs of grazing farmed animals on U.S. public lands is greater than greenbacks—even 100 million greenbacks. Public lands are being managed/damaged for the sake of low-cost flesh. Furthermore, for those who hike many miles into a wilderness that is presumably preserved for the public, it is an outrage to come upon a herd of bovines standing in a trampled, manure-riddled wasteland: Bovines "reduce aesthetics with their fecal matter, with the trampling of vegetation, and with their mere presence" ("Cattle Grazing"). Unfortunately, the vast majority of taxpayers are not aware of the environmental impact of ranching on public lands—or the financial losses associated with this environmentally disastrous government subsidy.

- Grazing
    - Endangers fragile species
    - Alters ecosystems
- Walking
    - Harm native plants, root systems, and waterfowl nests
    - Harm soils and streambeds
- Defecating
    - Cause land and water
    - Cause eutrophication (dead zones)

(openclipart.org)

SI.7 Cattle Harm Ecosystems

Ranchers have a very strong lobby; they woo legislators and affect policies. But what ranchers prefer is not generally in the interest of wildlife, ecosystems, or the vast majority of U.S. citizens. Ranchers extirpated the swift fox from the Great Plains, then blocked reintroduction until they were sure that swift fox conservation efforts would not harm their economic interests ("Swift Fox"). Those invested in animal agriculture are now suing the U.S. government in the hope of reversing a recent decision to restore bison to Montana public lands (Preso 11). Animal agriculture "remains the leading stodgy opponent to otherwise popular efforts to reintroduce species like the Mexican gray wolf in Arizona and New Mexico ("Grazing").

Environmentalists are usually keen to eliminate nonnative species and reintroduce native species: Bovines are nonnative; they harm native wildlife and plants and destroy local ecosystems. Buying beef supports and encourages ranching, including grazing on public lands. Do environmentalists wish to support millions of nonnatives—bovines (and pigs and chickens)—at the expense of local ecosystems and native species?

Future Trends

The above environmental problems caused by animal agriculture are just the tip of the grain pile. In developing countries "meat production increased more than 450 percent between 1980 and 2010" (Reynolds and Nierenberg 13). In China, for example, where traditional diets typically contain very small amounts of flesh (and no dairy products), "consumption of meat—pork, beef, poultry, and mutton—has climbed several fold [since 1978], pushing China's total meat consumption far above that of the United States" Per person, U.S. flesh consumption is still far greater. (Figure 1.18) (Brown 187). Dairy products, recognized by the Chinese people for thousands of years for what they are—the nursing milk of *another* species—were not consumed for centuries, but with the influx of Western habits, cheese and milk are increasingly popular ("India, China"). It is ominously clear that trends in developing nations indicate a massive increase in the consumption of animal products, currently growing by 3 percent each year, leading to a 35 percent increase by 2015 and doubled demand by 2050 (Cassuto 7; Hawthorne 40), or perhaps by 2030 (Reynolds and Nierenberg 13). Developed countries can mitigate some of these effects if citizens adopt a vegan diet. Not only will such a change directly reduce

Source: USDA

FIGURE 1.18 Meat Consumption in China Compared with U.S. Consumption, 1960–2012

the environmental impact of our dietary choice, but we will also begin to change the dietary example that we set for other nations.

"Sustainable," "Grass Fed," "Organic," and "Local"

Even when animal products are labeled "sustainable," "humane," "local," "grass-fed," "organic," "free-range," and/or "cruelty free," choosing to consume animal products (rather than grains and greens directly) significantly increases our environmental footprint. A diet that includes animal products is a diet laden with GHGE, water pollution, freshwater depletion, deforestation, soil damage, and the destruction of wildlife. Every product stemming from animal agriculture contributes mightily to GHGE. Every farmed animal poops. When fed grains, farmed animals not only consume vital foods that hungry humans might otherwise eat, but also tremendous amounts of freshwater. Every farmed animal requires land, altering and threatening landscapes and ecosystems. No matter what comfort-the-consumer label is placed on the final product, purchasing meat, dairy, and eggs demonstrates reckless disregard for the environment and is unconscionable when vegan foods are available and affordable—and bulk vegan staples are generally less expensive than animal products. (See chapter 3 for a cost comparison.)

- Greenhouse gas emissions

- Water pollution and dead zones

- Deforestation

- Extirpation and extinctions

- Soil degradation

- Freshwater depletion

(clker.com)

s1.8 The Affects of Animal Agriculture

Conclusion

In 2008 the United Nations released *Livestock's Long Shadow* (Steinfeld et al.), a groundbreaking report identifying animal agriculture as the *number one* contributor to climate change, clearly labeling animal agriculture as environmentally disastrous in a host of ways. Unfortunately, the authors of this otherwise thorough and respectable study shied away from the obvious conclusion—if we intend to protect the environment, we must stop producing and consuming animals and animal products.

The authors of *Livestock's Long Shadow* are not alone. Despite the fact that animal agriculture burns up deadly fossil fuels, massive quantities of luscious grain, vital freshwater, and diverse forests complete with precious ecosystems, leaving us with a remarkable quantity of GHGE, polluted and depleted freshwater reserves, dead zones, diminished wildlands, environmentalists tend to be unaccountably silent on the topic of dietary choice. Rather than alter meal plans—and ask others to do the same— environmentalists have preferred to urge that we use less water on lawns and in the shower, choose gas-efficient cars and fluorescent bulbs, and recycle cardboard and glass. While these are important measures, each is comparatively irrelevant when compared with the environmental benefits of shifting to a vegan diet. Eating plants and grains without cycling them through farmed animals is by far the most important change we can make on behalf of the environment. If you *sincerely* care about the planet, and/or about animals, and/or about humanity, can you offer even one *legitimate* reason to continue consuming animal products?

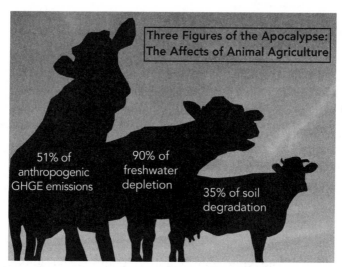

si.9 Three Figures of the Apocalypse: The Affects of Animal Agriculture

References

"A Broken Food Chain." *AWI Quarterly* 58.1 (Winter 2009): 18.

"Acid Rain: Effects Felt through the Food Chain." *National Geographic.* Accessed 27 May 2014. <http://environment.nationalgeographic.com/environment/global-warming/acid-rain-overview/>.

"Agriculture in Brazil." *Wikipedia.* Accessed 28 Nov. 2013. <http://en.wikipedia.org/wiki/Agriculture_in_Brazil>.

"Arsenic." *GreenFacts: Facts on Health and the Environment.* Accessed 28 Nov. 2013. <http://www.greenfacts.org/en/arsenic/index.htm>.

"Beef Cattle Production Statistics." *Statistics Brain.* Accessed 27 May 27 2013. <http://www.statisticbrain.com/beef-cattle-production-statistics/>.

Bernstein, Lenny, et al. *Climate Change 2007: Synthesis Report: An Assessment of the Intergovernmental Panel on Climate Change.* 12–17 Nov. 2007. Accessed 14 Nov. 2013. <http://www.ipcc.ch/pdf/assessment-report/ar4/syr/ar4_syr.pdf>.

"Big Manure Spill at Dairy Flows into Snohomish River." *Seattletimes.com.* 13 April 2010. Accessed 28 Dec. 2011. <http://seattletimes.nwsource.com/html/local news/2011597899_manure14m.html>.

"BLM and Forest Service Announce 2011 Grazing Fee." *Bureau of Land Management.* Washington Office Division of Public Affairs. 31 Jan 2011. Accessed 28 Dec. 2011. <http://www.blm.gov/wo/st/en/info/newsroom/2011/january/NR_01_31_2011.html>.

Brown, Lester R. *Plan B 3.0: Mobilizing to Save Civilization.* New York: W. W. Norton, 2008.

Butler, Rhett A. "Causes of Deforestation in the Brazilian Amazon." *Mongabay.com.* Accessed 13 Apr. 2013. <http://www.mongabay.com/brazil.html>.

Calverd, Alan. "A Radical Approach to Kyoto: Abandoning Meat is a Simple, Reversible Experiment." *Physics World* July 2005: 56. Quoted by David Reid in "Cut Global Warming by Becoming Vegetarian." *EurekAlert!* 6 July 2005. Accessed 22 Nov. 2013. <http://www.eurekalert.org/pub_releases/2005-07/iop-cgw070505.php>.

"Cattle and Beef." *USDA Economic Research Service.* United States Department of Agriculture. Accessed 15 Nov. 2012. <http://www.ers.usda.gov/topics/animal-products/cattle-beef/statistics-information.aspx>.

"Cattle Grazing on Public Lands: The Hard Fought Battle in the Southwestern United States." 5 Oct. 1996. *Thebeckoning.com.* Accessed 27 Dec. 2011. <http://www.thebeckoning.com/environment/cattle/grazing.html>.

Cassuto, David N. "The CAFO Hothouse: Climate Change, Industrial Agriculture, and the Law." Animals and Society Institute Policy Paper, 2010.

Chadwick, Douglas H. "Nose on the Range." *All Animals* January/February 2012: 22–27.

"Changes in the Size and Location of U.S. Dairy Farms." *Profits, Costs, and the Changing Structure of Dairy Farming.* USDA Economic Research Service. Economic Research Report ERR-47. September 2007. 2–4. Accessed 20 Feb. 2012. <http://www.ers.usda.gov/ersDownloadHandler.ashx?file=/media/430528/err47b_1_.pdf>.

"Chapter 6: Chicken Statistics." *How to Do Animal Rights.* Accessed 28 Nov. 2013. <http://www.animalethics.org.uk/i-ch7-2-chickens.html>.

Chastain, John P., and James J. Camberato. "Dairy Manure Production and Nutrient Content." Chapter 3a in "Confined Animal Manure Managers Certification Program Manual B Dairy Version 1." Clemson University Cooperative Extension Service. Clemson University. 2011. Accessed 28 Dec. 2011. <http://www.clemson.edu/extension/livestock/camm/camm_files/dairy/dch3a_04.pdf>.

"Clean Energy: Greenhouse Gas Equivalencies Calculator." *United States Environmental Protection Agency.* Accessed 14 Nov. 2013. <http://www.epa.gov/cleanenergy/energy-resources/calculator.html>.

"Climate Change 2001: Working Group I: The Scientific Basis: 4.2.1.2 Nitrous oxide (N2O)." *IPCC Third Assessment Report, Climate Change 2001, Complete Online Versions.* GRID Arendal. Accessed 14 Nov. 2013. <http://www.grida.no/publications/other/ipcc_tar/?src=/climate/ipcc_tar/wg1/134.htm#4211>.

Comerford, John W., George L. Greaser, H. Louis Moore, and Jayson K. Harper. "Agricultural Alternatives: Feeding Beef Cattle." Pennsylvania State University. 2001. Accessed 26 Dec. 2011. <http://pubs.cas.psu.edu/freepubs/pdfs/ua298.pdf>.

"Counting Chickens." *Economist Online.* 27 July 2011. Accessed 27 Dec. 2011. <http://www.economist.com/blogs/dailychart/2011/07/global-livestock-counts>.

Dawn, Karen. *Thanking the Monkey: Rethinking the Way We Treat Animals.* New York: Harper, 2008.

"Deforestation in Costa Rica." *Wikipedia.* Accessed 13 Apr. 2013. <http://en.wikipedia.org/wiki/Deforestation_in_Costa_Rica>.

"Deforestation: The Leading Cause of CO2 Emissions." *Global Warming Science.* Accessed 13 Apr. 2013. <http://www.appinsys.com/globalwarming/deforestation.htm>.

Dobson, Cathy. "Farm Manure Spill Contained." *London Free Press* 23 Nov. 2011. Accessed 28 Dec. 2011. <http://www.lfpress.com/news/london/2011/11/22/19003381.html>.

Driscoll, C. T., et al "Acid Rain Revisited: Advances in Scientific Understanding since the Passage of the 1970 and 1990 Clean Air Act Amendments." Hanover: Hubbard Brook Research Foundation, 2001. Accessed 27 Dec. 2011. <http://www.hubbardbrook.org/6-12_education/Glossary/AcidRain.pdf>.

"Ecological Footprint: What Is the Ecological Footprint?" *The Sustainable Scale Project.* Accessed 31 Jan. 2013. <http://www.sustainablescale.org/conceptualframework/understandingscale/measuringscale/ecologicalfootprint.aspx>.

"Effects of Changing the Carbon Cycle." *Features: The Carbon Cycle.* NASA Earth Observatory. Accessed 15 Nov. 2012. <http://earthobservatory.nasa.gov/Features/CarbonCycle/page5.php>.

"Enteric Fermentation Mitigation." *Center for Climate and Energy Solutions.* Accessed 13 Apr. 2013. <http://www.c2es.org/technology/factsheet/EntericFermentation>.

"Expert Assessment: Ocean Acidification May Increase 170 Percent This Century." *e! Science News.* Accessed 14 Nov. 2013. <http://esciencenews.com/articles/2013/11/13/expert.assessment.ocean.acidification.may.increase.170.percent.century>.

Fanelli, Daniele. "Meat Is Murder on the Environment." *New Scientist* 18 July 2007.

"Feedlot Industry—AS Flashcards." *Quizlet.* Accessed 26 Dec. 2011. <http://quizlet.com/1939629/feedlot-industry-as-flash-cards/>.

"Final Report: Total Maximum Daily Load Report for 58 Bacteria Impaired Waters in New Hampshire." State of New Hampshire, Department of Environmental Services, Water Division, Watershed Management Bureau, Aug. 2011. Accessed 26 Dec. 2011. <http://des.nh.gov/organization/divisions/water/wmb/tmdl/documents/finalreport58bacteriatmdls.pdf>.

"Fiscal Costs of Federal Public Lands Livestock Grazing." *WildEarth Guardians.* Sagebrushsea.org. Accessed 27 Dec. 2011. <http://www.sagebrushsea.org/pdf/factsheet_Grazing_Fiscal_Costs.pdf>.

"Forest Holocaust." *National Geographic: Eye in the Sky.* Accessed 13 Apr. 2013. <http://www.nationalgeographic.com/eye/deforestation/effect.html>.

Fox, Camilla H. "Predators in Peril: The Federal Government's War on Wildlife." *Indiana Coyote Rescue Center Newsletter* Winter 2009. Accessed 16 Nov. 2012. <http://www.coyoterescue.org/newsletters/winter2009.html>.

"Future Technology." *Future Electro Tech.* 23 Oct. 2008. Accessed 28 Dec. 2011. <http://futureelectrotech.blogspot.com/2008/10/agriculture-is-responsible-for.html>.

Gaechter, Lacey. "Environmental Implications of Modern Animal Agriculture: Save the Planet with Your Fork." Accessed 27 Dec. 2011. <http://www.veganoutreach. org/whyvegan/gaechter.html>.

Gluckman, Matt. "EPA Regional CAFO Waste Issues." *Epa.gov.* Accessed 27 Dec. 2011. <http://www.epa.gov/ncer/publications/workshop/pdf/gluckman_ region582007.pdf>.

Goodland, Robert, and Jeff Anhang. "Livestock and Climate Change." *World Watch* Nov./Dec. 2009: 10–19. Accessed 10 April 2013. <http://www.worldwatch.org/ files/pdf/Livestock%20and%20Climate%20Change.pdf>.

"Grain Harvest Sets Record, But Supplies Still Tight." *Worldwatch Institute.* Accessed 15 Nov. 2012. <http://www.worldwatch.org/node/5539>.

Grant, Rick, and Paul J. Kononoff. "Feeding to Maximize Milk Protein and Fat Yields." *NebGuide* Feb. 2007. Accessed 28 Dec. 2011. <http://www.ianrpubs. unl.edu/pages/publicationD.jsp?publicationId=703>.

"Grazing." *Center for Biological Diversity.* Accessed 27 Dec. 2011. <http://www.bio logicaldiversity.org/programs/public_lands/grazing/>.

"Greenhouse Gas Emissions from Animal Agriculture." *Humane Society of the United States.* Accessed 31 Jan. 2013. <http://www.humanesociety.org/ assets/pdfs/farm/hsus-fact-sheet-greenhouse-gas-emissions-from-animal-agriculture.pdf>.

Halden, Rolf U. and Kellogg J. Schwab. "Environmental Impact of Industrial Farm Animal Production: A Report of the Pew Commission on Industrial Farm Animal Production." Accessed 27 May 2014. <http://www.ncifap.org/_ images/212-4_EnvImpact_tc_Final.pdf>.

Halley, John. Personal Communication by e-mail, 1 Apr. 2014.

Hawthorne, Mark. "Planet in Peril." *VegNews* March–April 2012: 34–41.

Hein, R., P. J. Crutzen, and M. Heimann. "An Inverse Modeling Approach to Investigate the Global Atmospheric Methane Cycle." *Global Biogeochem Cycles* 11: 43–76, referenced in "Table 4.2: Estimates of the Global Methane Budget." *IPCC Third Assessment Report, Climate Change 2001, Complete Online Versions. GRID Arendal.* Accessed 14 Nov. 2013. <http://www.grida.no/publications/ other/ipcc_tar/?src=/climate/ipcc_tar/wg1/134.htm#4211>.

Heineck, Ken. "Grayson County Drinking Water Safe, after Manure Spill in Fries." 31 May 2011. *WSLS 10.* Accessed 28 Dec. 2011. <http://www.wsls.com/story/20817100/ grayson-county-drinking-water-safe-after-manure-spill-in-fries>.

Honnold, Douglas and Jenny Harbine. Letter to Governor Brian Schweitzer. 25 Feb. 2011. Accessed 27 May 2014. <http://earthjustice.org/sites/default/files/ Schweitzer-Letter-Final.pdf>.

"Hormones Used in Livestock Production." *Office of the Auditor General of Canada.* Accessed 29 Nov. 2013. <http://www.oag-bvg.gc.ca/internet/English/ pet_203_e_28939.html>.

"How Many People Can an Entire Cow Feed?" *Wiki Answers.* Accessed 26 Dec. 2011. <http://wiki.answers.com/Q/How_many_people_can_an_entire_cow_feed>.

"How Much Feed Will My Cow Eat—Frequently Asked Questions." Alberta Agriculture and Rural Development. Accessed 1 Nov. 2013. <http://www1.agric. gov.ab.ca/%24department/deptdocs.nsf/all/faq7811>.

"How Much Meat Is There on a Cow?" *Ask MetaFilter.* Accessed 26 Dec. 2011. <http://ask.metafilter.com/27259/How-much-meat-is-there-on-a-cow>.

"How Much Meat Will One Cow Provide?" *Yahoo! Answers.* Accessed 26 Dec. 2011. <http://answers.yahoo.com/question/index?qid=20061019172556AAtjvcG>.

"How Much Water Goes Into a Burger? Studies Find Different Answers." *Wall Street Journal* 11 Jan. 2008. Accessed 23 Nov. 2011. <http://online.wsj.com/arti cle/SB120001666638282817.html>.

"Howling for Justice: Blogging for the Grey Wolf." Accessed 27 May 2014. <http:// howlingforjustice.files.wordpress.com/2011/02/wolves-running.jpg >.

"Hygiene: Liquid and Solid Waste Disposal." *Slaughterhouse Cleaning and Sanitation.* FAO Corporate Document Repository. Accessed 6 Feb. 2012. <http://www.fao. org/docrep/003/x6557e/X6557E02.htm>.

"India, China to Boost Global Milk Production: Tetra Pak." *Economic Times* 11 July 2011. Accessed 31 Jan. 2013. <http://articles.economictimes.indiatimes.com/2011-07- 11/news/29761312_1_flavoured-milk-liquid-dairy-products-tetra-pak>.

"Intensive Farming Can Make Food Security Worse for the World's Hungry." *WSPA International* 19 Oct. 2012. Accessed 15 Nov. 2012. <http://www. wspa-international.org/latestnews/2012/food-security-worlds-hunger.aspx>.

Kaufman, Stephen R. *Guided by the Faith of Christ: Seeking to Stop Violence and Scapegoating.* Cleveland: Vegetarian Advocates, 2008.

Kaufman, Stephen R, and Nathan Braun. *Good News for All Creation: Vegetarianism as Christian Stewardship.* Cleveland: Vegetarian Advocates, 2004.

Kemmerer, Lisa. *Sister Species: Women, Animals, and Social Justice.* "Introduction." Urbana: U of Illinois, 2011. 1-43

Kinsman, R., F. D. Sauer, H. A. Jackson, and M. S. Wolynetz. "Methane and Carbon Dioxide Emissions from Dairy Cows in a Full Lactation Monitored over a Six-month Period." *Journal of Dairy Science* 78 (Dec. 1995): 2760–66. Accessed 1 Nov. 2013.

"Land Degradation."*Global Change.* 4 Jan. 2010. Accessed 26 Dec. 2011. <http:// www.globalchange.umich.edu/globalchange2/current/lectures/land_deg/ land_deg.html>.

"Land Use Overview." *Ag 101.* United States Environmental Protection Agency. 9 Sept. 2009. Accessed 26 Dec. 2011. <http://www.epa.gov/agriculture/ag101/lan duse.html>.

Lange, Karen E. "Coyotes among Us." *All Animals* May/June 2012: 14–19.

Leopold, Aldo. "Wanted—National Forest Game Refuges." *Bulletin of the American Game Protective Association* 9.1 (1920). Available online at *The Aldo Leopold Archive.* Accessed 2 Feb. 2013. <http://digicoll.library.wisc.edu/cgi-bin/AldoLeopold/ AldoLeopold-idx?type=turn&entity=AldoLeopold.ALReprints.p0451&id= AldoLeopold.ALReprints&isize=text>.

Litke, David W. "Historical Water-Quality Data for the High Plains Regional Ground-Water Study Area in Colorado, Kansas, Nebraska, New Mexico, Oklahoma, South Dakota, Texas, and Wyoming, 1930-98" *USGS: Water Resources of Colorado*. Accessed 27 May 2014. <http://pubs.usgs.gov/wri/wri00-4254/>.

Steinfeld, Henning, et al. *Livestock's Long Shadow: Environmental Issues and Options*. Rome: Food and Agriculture Organization of the United Nations, 2006.

"Major Crops Grown in the United States." *Ag 101*. United States Environmental Protection Agency. 10 Sept. 2009. Accessed 27 May 2014. <http://www.epa.gov/agriculture/ag101/cropmajor.html>.

"Manure Spill Clears Out Black River." *Rodale Institute*. 15 Aug. 2005. Accessed 28 Dec. 2011. <http://newfarm.rodaleinstitute.org/news/2005/0805/081205/spill.shtml>.

"Manure Spill over Mile Long Closes Md. Road." *NBC4 Washington*. 19 Sept. 2011. Accessed 27 Dec. 2011. <http://www.nbcwashington.com/news/local/Manure-Spill-Over-Mile-Long-Closes-Md-Road-130105248.html>.

Matheny, Gaverick. "Least Harm: A Defense of Vegetarianism from Steven Davis's Omnivorous Proposal." *Journal of Agricultural and Environmental Ethics* 16.5 (2003): 505–11. Accessed 27 Dec. 2011. <http://www.veganoutreach.org/enewsletter/matheny.html>.

McManus, John. "Years of Work Pay Off for Klamath River Restoration." *Earthjustice* Winter 2010/2011: 15.

"Methane vs. CO2 Global Warming Potential." *Global Warning Forecasts*. Accessed 28 Dec. 2011. <http://www.global-warming-forecasts.com/methane-carbon-dioxide.php>.

Mohr, Noam. "A New Global Warming Strategy: How Environmentalists are Overlooking Vegetarianism as the Most Effective Tool against Climate Change in Our Lifetimes." *EarthSave*. Accessed 31 Jan. 2013. <http://www.earthsave.org/globalwarming.htm>.

Naze, Dale W., John Dhuyvetter, and Chip Poland. *Value-Based Beef Cattle Production*. North Dakota State University. Jan. 1999. Accessed 26 Dec. 2011. <http://library.ndsu.edu/tools/dspace/load/?file=/repository/bitstream/handle/10365/9238/AS-1163-1999.pdf?sequence=2>.

Nicholoff, Sharon. "The Rest of Our Wildlife: A Report on the Status and Future of Wyoming's Terrestrial Nongame." *Wyoming Wildlife* March 2004: 15–34

"Nitrous Oxide Emissions: Greenhouse Gas Emissions." *United States Environmental Protection Agency: Climate Change*. Accessed 31 Jan. 2013. <http://epa.gov/climatechange/ghgemissions/gases/n2o.html>.

"NOAA Study Shows Nitrous Oxide Now Top Ozone-Depleting Emission." *NOAA National Oceanic and Atmospheric Administration*. 27 Aug. 2009. Accessed 31 Jan. 2013. <http://www.noaanews.noaa.gov/stories2009/20090827_ozone.html>.

Oppenlander, Richard A. *Comfortably Unaware: Global Depletion and Food Responsibility . . . What You Choose to Eat Is Killing our Planet*. Minneapolis: Langdon Street, 2011.

"Oregon, Washington and Idaho seek to resume killing sea lions at Bonneville Dam." *OregonLive.com.* Accessed 9 May 2012. <http://www.oregonlive.com/pacific-northwest-news/index.ssf/2011/08/oregon_washington_and_idaho_seek_to_resume_killing_sea_lions_at_bonneville_dam.html>.

O'Toole, Randal. "Audit of the U.S.D.A. Animal Damage Control Program." *Thoreau Institute.* Accessed 16 Nov. 2012. <http://www.ti.org/adcreport.html>.

Pepino, Chrissy. "National: Too Much Drought, Too Little Water in West." On the Frontlines: Cases, Issues, Victories. *EarthJustice* Spring 2014: 6.

Perlman, David. "Scientists Alarmed by Ocean Dead-Zone Growth." *SFGate.* 15 Aug. 2008. Accessed 28 July 2013. <http://www.sfgate.com/green/article/Scientists-alarmed-by-ocean-dead-zone-growth-3200041.php>.

"Pig Facts and Pig Trivia," HubPages. Accessed 27 May 2014. <http://drbj.hub pages.com/hub/Pig-Facts>

Pimm, Stuart L., and Peter Raven. "Extinction by Numbers." *Nature* 24 Feb. 2000: 843–45.

Preso, Tim. "Wild Bison v Cattle." *EarthJustice* Summer 2012: 11.

"Public Lands Ranching: The Ecological Costs of Public Lands Ranching." *Western Watersheds Project.* 2010. Accessed 28 Dec. 2011. <http://westernwatersheds. org/issues/public-lands-ranching?gclid=CIGqkcaruaoCFQ5lhwod V2YG4w>.

"Rainforest Facts." *Raintree.com.* 20 Mar. 2010. Accessed 28 Dec. 2011. <http:// www.rain-tree.com/facts.htm>.

Regan, Tom. *The Case for Animal Rights.* New York: Routledge, 1984.

Reynolds, Laura, and Danielle Nierenberg. *Innovations in Sustainable Agriculture: Supporting Climate-Friendly Food Production.* Worldwatch Report 188. Washington, DC: Worldwatch Institute, 2012.

Richmond, Bill. "State Representation Candidate." *News-Gazette* 12 Aug. 2008. Accessed 26 Dec. 2011. <http://www.winchesternewsgazette.com/articles/2008/08/13/news/news2.txt>.

Ritz, Casey W., and William C. Merka. "Maximizing Poultry Manure Utilization through Nutrient Management Planning." *University of Georgia College of Agricultural and Environmental Sciences Cooperative Extension Service.* Accessed 14 Nov. 2013. <http://www.poultry.uga.edu/pdffiles/manureutiliz.pdf>.

"The Role of Livestock in Climate Change." *Food and Agriculture Organization of the United Nations: Livestock, Environment, and Development.* Accessed 27 May 2013. <http://www.fao.org/agriculture/lead/themeso/climate/en/>.

Sachs, Jeffrey D. "The Promise of the Blue Revolution." *Scientific American* July 2007: 37–38.

Schmitt, Angie. "The American Cities with the Most Growth in Car-free Households." *Greater Greater Washington: The Washington, DC Area is Great. But it could be Greater.* Accessed 27 May 2014. <http://greatergreaterwashington.org/post/21444/the-american-cities-with-the-most-growth-in-car-free-households/>.

Schwartz, Richard H. *Judaism and Vegetarianism.* New York: Lantern, 2001.

"Separate Hamilton County Fish Kills Total More Than 30,000 Fish." Iowa Department of Natural Resources. 10 Aug. 2011. Accessed 27 Dec. 2011. <http://www.iowadnr.gov/home/ctl/detail/mid/2822/itemid/363>.

"Swift Fox." *Blue Planet Biomes.* Accessed 2 Feb. 2013. <http://www.blueplanetbiomes.org/swift_fox.htm>.

"Table G-2: Animals Euthanized or Killed by Wildlife Services—FY 2011." United States Department of Agriculture: Animal and Plant Health Inspection Service: Wildlife Damage Management. Accessed 16 Nov. 2012. <http://www.aphis.usda.gov/wildlife_damage/prog_data/2011_prog_data/PDR_G/Basic_Tables_PDR_G/Table_G-2_Euth-Killed.pdf>.

"The Top 5 Most Common Types of Livestock Feed Used Worldwide." *Top 5 of Anything: Farming & Agriculture Statistics.* Accessed Nov. 15, 2012. <http://top50fanything.com/index.php?h=2aadbfa5>.

Tomer, Adie. "Transit Access and Zero-Vehicle Households." *Metropolitan Policy Program at Brookings: Metropolitan Infrastructure Initiative Series and Metropolitan Opportunity Series.* Accessed 27 May 2014. <http://www.brookings.edu/~/media/research/files/papers/2011/8/18%20transportation%20tomer%20puentes/0818_transportation_tomer.pdf>.

Turner, Tom. "U.S. Senate, House Take Aim at the Wolves: Many Other Species at Risk as Well." *In Brief* Spring 2011: 10–11.

Vaquer-Sunyer, Raquel, and Carlos M. Duartel. "Thresholds of Hypoxia for Marine Biodiversity." *Proceedings of the National Academy of the Sciences of the United States of America* 105.40 (2008): 15452–57. <http://www.pnas.org/content/105/40/15452.full>.

Voisinet, B. D., T. Grandin, J. D. Tatum, S. F. O'Connor, and J. J. Struthers. "Feedlot Cattle with Calm Temperaments Have Higher Average Daily Gains Than Cattle with Excitable Temperaments." *Journal of Animal Science* 75 (1997): 892–96. Accessed 1 Nov. 2013. <http://www.grandin.com/references/gains.html.>

Ward, D., and K. McKague. "Water Requirements of Livestock." *Ontario Ministry of Agriculture, Food, and Rural Affairs.* May 2007. Accessed 1 Nov. 2013. <http://www.omafra.gov.on.ca/english/engineer/facts/07-023.pdf>.

Wardle, Tony. "Diet of Disaster." *Viva!* Accessed 28 Dec. 2011. <http://www.viva.org.uk/hot/dietofdisaster/>.

"What Are the Main Sources of Nitrous Oxide Emissions?" *What's Your Impact on Climate Change?* Accessed 28 Dec. 2011. <http://www.whatsyourimpact.eu.org/n2o-sources.php>.

Whitty, Julia. "The End of a Myth." *On Earth* Spring 2012: 39–43.

"Wildlife Killed." *WildEarth Guardians.* Accessed 26 Dec. 2011. <http://www.wildearthguardians.org/site/PageServer?pagename=priorities_wildlife_war_wildlife_killed_table>.

"Wildlife Services Under Fire, Releases Annual Kill Numbers: Overall Kills Declined, but Native Carnivores Endured Increased Mortality." *WildEarth*

Guardians 8 May 2012. Accessed 16 Nov. 2012. <http://www.wildearthguard ians.org/site/News2?page=NewsArticle&id=7639>.

Worm, Kally. "Groundwater Drawdown." *Water Is Life* Spring 2004. Accessed 26 Dec. 2011. <http://academic.evergreen.edu/g/grossmaz/WORMKA/>.

ZoBell, Dale R., Craig Burrell, and Clell Bagley. "Raising Beef Cattle on a Few Acres." Utah State University Extension. Sept. 1999. Accessed 26 Feb. 2012. <http://digitalcommons.usu.edu/cgi/viewcontent.cgi?article=1024&context=e xtension_histall>.

2

A Fishy Business

[The] Ocean is the largest wilderness on Earth, home to wildlife in staggering mul-
tispecies aggregations, and with a lineage of life three billion years older than anything above sea level. Its three-dimensional realm comprises 99 percent of all habitable
space and is so embedded with life as to be largely comprised of life, with an ounce of
seawater home to as many as 30 billion microorganisms—and counting.
—WHITTY 40

OCEANS COVER THE majority of the planet and are home to a vast quantity of diverse yet interconnected ecosystems. The volume of living space
provided by the sea is 168 times greater than that provided by the earth's
landscapes (Clark et al. 5). Given the wealth of creatures living in the seas,
as well as those in lakes, swamps, and rivers, and given the much-touted
health aspects of aquatic flesh and the environmental nightmare linked with
animal agriculture, should we turn to a diet of pollock, shrimp, and salmon?

The "Healthy" Flesh

In the 18th and 19th centuries, a mercury wash was used to produce felt
hats. In the process workers were exposed to and absorbed bits of the
substance and developed mercury poisoning. As a consequence, those
who were employed in the felt hat industry often stumbled about "in
a confused state with slurred speech and trembling hands" and "were
sometimes mistaken for drunks" ("Mad as a Hatter"). Mercury poisoning "attacks the nervous system, causing drooling, hair loss, uncontrol-
lable muscle twitching, a lurching gait, and difficulties in talking and
thinking" ("Mad as a Hatter"). From this comes the term made famous in
Lewis Carroll's *Alice in Wonderland,* "mad as a hatter."

Between 1973 and 1997, fish consumption rose from 45 to 91 million
metric tons (Delgado 1). The American Heart Association recommends

eating fish at least twice a week for heart health ("Omega-3 Fatty Acids"). The National Healthy Mothers, Healthy Babies Coalition accepted "thousands of dollars from the fishing industry" to promote a recommendation that pregnant women eat "at least 12 ounces of fish per week" ("Fishy Recommendations" 23). Fish flesh is touted as "healthy meat" in comparison with the flesh of terrestrial animals. Omega-3 fatty acids found in fish are credited with helping everything from heart disease to diabetes, but fish flesh also contains deadly mercury (as well as "dioxins . . . and polychlorinated biphenyls"—PCBs) ("Omega-3 Fatty Acids"). Longer-living predator fish such as tuna, marlin, shark, mackerel, and swordfish "can have mercury concentrations that are hundreds or thousands of times, possibly even a million times, greater than concentrations in the water in which they swim" (Smith and Lourie 151).

Mercury, one of the most toxic substances in our environment, crosses both "the blood-brain barrier and . . . the placental barrier" (Smith and Lourie 140, 154). Once inside, mercury concentrates in major organs, including the heart, liver, kidneys, and brain, where it dissolves "neurons in certain parts of the brain . . . leading to various nervous system disorders" (Smith and Lourie 155, 146). Mercury poisoning

> can cause permanent brain damage, central nervous system disorders, memory loss, heart disease, kidney failure, liver damage, cancer, loss of vision, loss of sensation and tremors. It is also among the suspected "endocrine disruptors," which do damage to the reproductive and hormonal development of fetuses and infants. Some studies also suggest that mercury may be linked to neurological diseases, such as multiple sclerosis, attention deficit disorder, and Parkinson's, but the evidence here remains somewhat inconclusive. (Smith and Lourie 145)

The type of mercury that fish carry, methyl mercury, is easily absorbed from the digestive tract, and once in the bloodstream, accumulates in the brain (Griesbauer). There is no "safe" level with regard to mercury. Even low doses of mercury over an extended period of time "cause serious physical and mental impairment" (Smith and Lourie 146). Mercury is a "potent neurotoxin that will kill you" (Smith and Lourie 146).

Presumably, when they promoted fish flesh for pregnant women, the National Healthy Mothers, Healthy Babies Coalition did not know that very young humans—including humans in the womb—are particularly susceptible to the ravaging effects of mercury, though even at the time this information was posted in almost every state, with "warnings to limit the amount of certain types of fish consumed by pregnant and nursing women due to mercury contamination" ("Mercury from Cement Kilns" 17). Bulletins around the world warn pregnant women to use caution if they choose to consume fish. Nonetheless, "one in six women of child-bearing age has elevated mercury levels," and this is yet more likely for women of indigenous coastal communities such as Canada's Arctic villages, where mercury blood levels for one in three women are above what is admitted to be dangerous (Smith and Lourie 157, 131,155).

While very young children are particularly susceptible to the terrifying effects of mercury, this poison is also damaging for adults. If you are a skilled angler who consumes what you catch, you will have "no trouble poisoning yourself"—consuming almost any predator fish is a good way to be poisoned (Smith and Lourie 137, 158). The U.S. Environmental Protection Agency recommendations for acceptable levels of mercury can easily be exceeded with just a handful of fishy meals from the grocery store. Three servings of tuna, for example, can double mercury blood levels in less than 48 hours. Author Bruce Lourie tested the effects of fish (and other poisons) on his own body. After eating 7 fishy servings in the course of three days, his blood mercury levels moved from roughly 3.5 to more than 8.5, well above levels that the U.S. government considers "safe" (despite the fact that no amount of mercury is safe for human beings) (Smith and Lourie 139). In the process, Lourie came to understand "how communities that depend on fish in their diets can quietly poison themselves"—like the people of Minamata, Japan, in 1956 when despite the fact that mercury levels in the water and in fish were significantly lower at that time (Smith and Lourie 138, 150).

Why is there so much mercury in oceans and fish? Industries—especially coal plants—pollute the environment with mercury, which flows into streams and lakes, then into seas. Pollutants such as mercury (as well as uranium, and—the classic—DDT), which living beings cannot break down, accumulate. As a result, these micro-particles build up in body tissues in the same way that plastics build up in the stomachs of sea birds who inadvertently swallow, but cannot process, plastics. Natural substances (proteins, carbohydrates, and fat) are broken down, absorbed, and utilized, but mercury cannot be broken down, absorbed, or utilized, so mercury accumulates. When these mercury-ridden sea creatures are eaten by predators, including humans, this poison accumulates in a process called "bioaccumulation" (Figure 2.1). When larger predators such as sharks, humans, or eagles eat many fish over the course of their lives, fish

When one fish eats another, indigestible matter (bones, teeth, etc.) is usually excreted, but some substances (like mercury) cannot be processed.

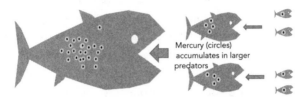

Mercury (circles) accumulates in larger predators

If a shark eats 33 smaller fish, each containing 2 mg of mercury, the shark absorbs—and retains—66 mg of mercury. In this way, toxic substances become concentrated in larger predators.

FIGURE 2.1 Bioaccumulation

who contain particles of mercury, uranium, and DDT, these larger predators store more and more noxious chemicals in their bodies.

Thanks to human pollution, all fish are contaminated with mercury—it's only a question of degree. Nonetheless, many consumers have been led to believe that fish is healthier than bacon or eggs. . . even as "global mercury levels continue to rise" (Smith and Lourie 158).

Omega fatty acids are readily available in plant foods (most notably flax and hemp seeds/oils). Plant foods do not contain mercury (or cholesterol). The catfish is out of the bag—those interested in human health will avoid eating fish.

What Is a Fish?

While informed people now worry about mercury levels in fish flesh, few are concerned about the suffering and loss of life that is inherent in fish consumption. It seems that these creatures of the sea suffer most from is simply being out of sight and out of mind—and from lacking the warm fuzzy bodies that tend to spark human empathy. Despite the fact that we cannot hear them cry out in pain, or see them grimace, they are sentient—and suffering.

Fish are remarkable creatures who can and do suffer from "pain, fear and stress" (Chandroo, Duncan, and Moccia). They are vertebrates with a complex nervous system. Anatomical, pharmacological, and behavioral data, as well as evolutionary evidence and neurophysiological analogies, demonstrate that fish suffer (Chandroo, Duncan, and Moccia; Rollin 31). It is "unthinkable that fish do not have pain receptors; they need them in order to survive" ("Problem with Fishing"). Fish have nerve endings designed to register pain, just like other vertebrates, and fish produce the same brain chemicals that humans produce to counter pain: enkephalins and endorphins (Balcombe 187).

Fish are vertebrates with a complex nervous system.

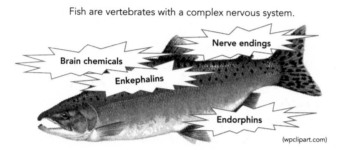

Nerve endings

Brain chemicals

Enkephalins

Endorphins

(wpclipart.com)

Anatomical, pharmacological, and behavioral data, as well as evolutionary evidence and neurophysiological analogies, indicate that fish are sentient.

S2.1 Do Fish Feel Pain?

Goldfish respond to pain "consciously, rather than simply reacting with a reflex" (Black). When subjected to pain, they act fearful, exhibiting "avoidance behavior," and such behavior "is cognitive—not reflexive" (Black). They become upset when frightened or harmed—when handled roughly—and hormone levels do not readjust for 48 hours (Balcombe 186–87).

[W]hen noxious substances were applied to the lips of trout, the fishes' heart rates increased, and they took longer to resume feeding. These fish also exhibited unusual behaviors after being harmed, including rocking from side to side while balanced on their pectoral fins, and rubbing their lips into the gravel and against the tank walls. Treatment with a pain suppressant significantly lowered these reactions. Other experiments have found that fish learn to avoid unpleasant stimuli such as electric shocks, and piercing of their lips by sharp hooks. (Balcombe 187)

Fish interact socially, learn, and remember what they learn. The brains of fish, like those of other sentient vertebrates, provide a means by which to avoid suffering, including a good memory and the ability to learn (Dionys de Leeuw 378). Fish are intelligent—intelligent enough to use tools (Balcombe 188). This is all the more extraordinary given their finger-free anatomy. In Australia's Great Barrier Reef, a blackspot tuskfish was filmed using his mouth to grasp and strike a clam on a rock to access the flesh inside (Brown). In Palau in Micronesia, an orange dotted tusk-fish was filmed digging clams, toting them to rocks, then tossing them against the hard surface: "impressive when you consider that fish don't have hands" (Viegas). Such discoveries give reason to pause when we consider how little we have explored the creatures of the watery depths—and we have already found of sea beings that use tools. "We really need to spend more time filming underwater to find out just how common tool use is in marine fish" (Brown). Fish tested in artificial waters have also proven quite capable of navigating complex mazes, purposefully avoiding areas where they have been previously threatened, and showing clear signs of fear as they approach areas where they remember having been previously frightened or harmed (Rodriguez).

Fish use long-term memories to survive in waters riddled with predators, and to navigate a complex social world (Balcombe 188). We now know that fish are able to recognize "shoal mates," acknowledge hierarchy, track relationships, and eavesdrop on others in their community (Balcombe 188).

> Gone (or at least obsolete) is the image of fish as drudging and dim-witted pea brains, driven largely by "instinct," with what little behavioral flexibility they possess being severely hampered by an infamous "three-second memory." . . . Now, fish are regarded as steeped in social intelligence, pursuing Machiavellian strategies of manipulation, punishment and reconciliation, exhibiting stable cultural traditions, and co-operating to inspect predators and catch food. (Laland 4)

We should have known that a lack of fuzzy fur didn't mean much: Although surprisingly few people empathize with fish, they are much like other animals (including human beings), complete with emotions such as fear, memories that can guide them through mazes, and complex brains that facilitate tool use for such things as breaking clams against rocks. Like chickens, cattle, and human beings, "fish are conscious of their existence" (Dionys de Leeuw 378).

"If we're going to use animals in experiments, and we're going to use animals as food, then it is really important to understand the consequences of our actions for those animals" (Black).

> When hauled out of their natural environment by a hook embedded in their delicate mouths, fish thrash, writhe, and struggle in an attempt to escape. They gasp desperately in the air, unable to breathe out of water, as they slowly begin to suffocate. Like any other animals, their struggles indicate an aversion to pain and a strong will to survive. ("Problem with Fishing")

The pain that fish experience on a hook is likely comparable to dentistry without novocain ("Problem with Fishing"). All told, evidence indicates that fishing inspires a great deal of fear and causes a lot of suffering.

Fish are similar to other animals in ways that are morally relevant: "Fish are sensitive to pain, have memory and are capable of learning, and are conscious" of their existence (Dionys de Leeuw 325). In these ways fish are no different from other species, "such as. . . amphibians, reptiles, birds, mammals, and humans" (Dionys de Leeuw 325). In short, fish "have all the relevant characteristics attributable to animals requiring humane treatment" (Dionys de Leeuw 389). Unfortunately, this understanding of fish has not caused so much as a blip in our fish-eating ways.

- Fish interact socially, learn, and remember what they learn.
- Fish navigate complex mazes.
- Fish avoid areas where they have been threatened or harmed, showing signs of fear when approaching such areas.

(wpclipart.com)

Because they are intelligent and sentient, fish ought to be protected by animal welfare laws.

s2.2 Who Are fish?

Industrialized Fishing

Endangered: likely to become extinct.

Threatened: likely to become endangered in the near future.[1]

In 2003, the Pew Oceans Commission warned that the world's seas are in a state of "silent collapse" ("A Look"). *New Scientist* recently reported that an estimated 80 percent of fisheries have reached or exceeded maximum yields (D. MacKenzie). We have "removed as much as 90 percent of the large predatory fish such as shark, swordfish and cod from the world's oceans" ("A Look"). In the Pacific Northwest, where fish were once so thick that it seemed possible to walk across rivers "on the backs of salmon," commercial fish populations are now a tiny fraction of these historic levels ("ESA Salmon Listings"): "A 1991 report by the American Fisheries Society indicated that 214 of about 400 stocks of salmon, steelhead, and sea-run cutthroat trout in the Northwest and California are at risk of extinction. The report also indicated that 106 are already extinct" ("Pacific Salmon"). Only four salmon species remain in Washington/Oregon's Columbia River—pink salmon have been extirpated, and the remaining species are threatened, except sockeye, listed as endangered ("Salmon and Steelhead"; "Endangered Species"). New England groundfish have also dwindled—"Simply put, they are overfished" (Lawlor 13). "Our taste for seafood destroys 100 million tons [91 million metric tons] of wildlife annually" ("A Look").

 **Our Taste for Fish**

(clker.com)

- 80% of fisheries have reached or exceeded maximum yields.
- 90% of large predatory fish (shark, swordfish, cod) may already be fished out.
- 91 million metric tons of wildlife are fished from the sea annually.

(clker.com)

Thanks to fishing, the world's oceans are in a state of "silent collapse." (Pew Oceans Commission)

s2.3 Our Taste for Fish

---

1. "http://www.nwf.org/Wildlife/Wildlife-Conservation/Endangered-Species-Act.aspx".

Yet this has failed to frighten our fingers out of the fish bowl. The world's fishing fleet continues to grow, determined to satisfy an exponentially growing human population's appetite for sea flesh ("Global Fisheries"). And global fish intake continues to grow, with fish constituting about 20 percent of animal protein intake, not counting the "unrecorded contribution of subsistence fisheries" ("State of World Fisheries"). When fishing productivity on the continental shelf declined (due to overfishing and habitat destruction), fishers simply moved to deeper waters (Devine, Baker, and Haedrich 29), pulling up a "hitherto-unexploited wealth of strange-looking fish on the slopes of the continental shelves, down to 1600 metres" (D. MacKenzie). Consequently, at least five species of deepwater exotic fish—roundnose and onion-eye grenadier, Greenland halibut, blue hake, spiny eel, and spinytail skate—"only caught since the 1970s—are now on the critically endangered list. . . . [M]any other species are likely to be similarly endangered and . . . there seems little hope of saving them" (D. MacKenzie).

Target fish (such as swordfish and tuna) tend to be large, yielding more flesh (and therefore more profits), but top predators are important to stable ecosystems: Long-term studies indicate that the disappearance of larger fish from an ecosystem destabilizes fish populations, causing a more general decline that disproportionally affects fish with slower maturity rates (Stenseth 825).

Our taste for fish flesh is also shaping evolution: Reducing large predators in relation to smaller fish, and reducing the sizes of all species that we are likely to pull from the water. Moreover, larger fish tend to take longer to reach sexual maturity. The orange roughy, for example, must live for at least twenty years in order to reproduce, and apparently lives as long as 150 years ("Rough Going"). Orange roughy, only fished commercially since the '70s, via deep-water trawling, and only as more conventional, easier-to-catch favorites were depleted, has taken a precipitous population plunge in waters off of Australia, New Zealand, and Namibia ("Rough Going"). It is virtually impossible to target large, slow-to-reproduce fish sustainably, and species recovery is difficult, even unlikely (Raloff 360).

Additionally, we are shaping fish maturity and reproduction. Some populations of large fish are now reaching reproductive maturity earlier. The "process of maturing at younger ages and smaller sizes, called juvenescence, is usually a compensatory response to diminishing numbers within a population of fishes" ("Red Snapper FAQs"). In a heavily fished area on the coast of Alabama, red snapper now mature at two years of age and about 13 inches. A few hundred miles away, off the coast of Louisiana,

> **Orange Roughy**
> (*Hoplostethus atlanticus*)

(Wikipedia)

- Lives at a depth of 180-1,800 m/590-5,900 ft.
- Lives in unregulated international waters.
- Lives ~ 150 years.
- Takes at least twenty years to reach sexual maturity.
- Has an extremely low reproduction rate.

We lack data necessary to accurately predict sustainable fishing quotas for this species.

For all of these reasons, the orange roughy has suffered a precipitous population decline in recent years.

s2.4 Orange Roughy (*Hoplostethus atlanticus*)

red snapper mature at six years of age and about 27 inches ("Red Snapper FAQs"). Apparently, the only way "large" fish survive nets and hooks long enough to reproduce is to reach sexual maturity when they are younger and smaller. This reduces the overall size of a species, a change that may well be irreversible (Stenseth 826).

Juvenescence is also affecting bluefin tuna ("Status Review Report"). Bluefin tuna average 6.5 feet (2 m) and about 550 pounds (250 kg)—but sometimes reach 10 feet (more than 3 meters) and weigh 1,500 pounds (679 kg) ("Atlantic Bluefin Tuna"; Roach). Bluefin tuna are also among the fastest of fish, "shooting through the water with their powerful, crescent-shaped tails up to 43 miles (70 km) per hour" ("Atlantic Bluefin Tuna"). They are warm-blooded, "a rare trait among fish," and bluefin

> **Bluefin Tuna**

(commons.wikimedia.org)

- One of the largest fish (2 m and 250 kg; sometimes as large as 3 m and 679 kg).
- One of the fastest fish, up to 70 km/hour.
- One of the strongest distance swimmers.
- One of very few warm-blooded fish species.
- Suffered a 90% decline in adult biomass in just three generations.
- Can sell for as much as $4,009 per lb/$8,856 per kg.

s2.5 Bluefin Tuna

tuna are also "among the most ambitiously migratory of all fish"—sometimes "swimming from North American to European waters several times a year" ("Atlantic Bluefin Tuna").

These extraordinary fish are "plundered by fishing vessels" as they travel to the Gulf of Mexico to spawn, and once they arrive, they are "decimated by longline fishing boats seeking other species" (Lawlor 12). With soaring prices and an increase in demand, numbers of large, sexually mature bluefin tuna have plummeted, leaving only smaller, younger fish (Figure 2.2) ("Atlantic Bluefin Tuna). These beautiful metallic-blue fish, the biggest fish among tunas, have been "hunted to the brink of extinction," and "if current trends continue, the species will soon be functionally extinct in the Pacific" (Harvey). Bluefin tuna are "at risk of collapse": They have suffered a "90% decline in adult biomass within three generations, the criterion used by the IUCN for defining populations as *Critically Endangered*," and this *despite* an in-place "recovery plan" (B. MacKenzie et al. 25). If bluefin tuna are going to survive, we must leave enough sexually mature fish in the water to carry this fish's genetic code into the future. Yet even as these magnificent fish are dangerously depleted, their price per pound soars: In January of 2013, a single 444-pound (201-kg) bluefin tuna sold at auction for $1.78 million in Tokyo's Tsukiji Fish Market ("Bluefin"). Any viable recovery plan must require leaving all bluefin tuna in the water. Period.

Not surprisingly, scientists studying the seas have predicted that "all commercial fish and seafood species will collapse by 2048" (Stockstad

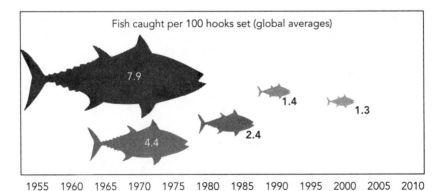

Included with permission: Stefan Kubicki, Save Our Seas Foundation ("Threat 1"), revised for print by John Halley

FIGURE 2.2 Global Decline in Bluefin Tuna

745). Are those consuming fish aware that they are eating sea life right out of the mouths of Steller sea lions, harbor seals, brown bears, and bald eagles? Recent news articles describe large numbers of dead animals washed up on beaches, "apparently" having starved to death. For example, we can expect to find more and more whales washed up who have "apparently" starved to death, more and more flocks of penguins who "apparently" cannot find enough to eat—including emperor penguins, whose primary source of food is now being stripped from seas to fatten farmed fish ("Apparently Starved Gray Whales"; Lehman; Bingham; Dawn 158). We are already eating our way through the first round of preferred substitutes (replacing fished-out species), such as pollock—a species critical to the diet of endangered Steller sea lions (Singer and Mason 115). As fisheries collapse, consumers are forced to shift to other dietary options, but what will sea lions eat?

Fishing Methods

The word "fishing" sometimes prompts images of a few kids down at the local fishin' hole, or perhaps a barefoot man pulling handwoven nets into a wooden boat, or a very long fishing line gracefully arching over the shoulder of a man in hip-high waders. But these are not the fishers, or the methods, that feed nations. And though fish are sometimes touted as earth-friendly in comparison with other forms of flesh, contemporary fishing is by no means sustainable or environmentally friendly (Sachs).

Commercial fishers use nets, hooks, and traps (or pots). Nets and hooks are the most common fishing methods—and the most environmentally damaging. Contemporary commercial fishing is patently unsustainable and environmentally devastating, in part because commercial fishing methods are indiscriminate: Hooks and nets attract and ensnare much more than the intended target fish.

Bycatch

Bycatch underscores the indiscriminate nature of industrial fishing. Bycatch, often referred to as "trash," are nontarget animals, including birds and marine mammals, some of which are listed as threatened or endangered. Whales, seals, sea lions, stingrays, porpoises, and sea turtles are drowned and/or pulled aboard incidentally in the quest for more marketable species. Twenty-two endangered seabird species are threatened by commercial fishing, nearly half of the earth's endangered seabird species. Some 300,00 seabirds perish on longlines annually when they are attracted to fishing lures and swallow longline hooks (http://cassiopaea. org/forum/index.php?topic=23793.0). Nontarget deep-sea fish are pulled up along with target species, their blood vessels and air bladders rupturing as they ascend. For obvious reasons, bycatch mortality rates are 90–100 percent (Clucas). "With a global marine mammal bycatch of several hundred thousand animals per year, [bycatch is] the primary threat to several endangered species of marine mammals" (Read, Drinker, and Northridge 168).

Bycatch endangers otherwise healthy species, and slows and even prevents recovery of decimated fish populations. River herring, demolished by trawlers, have failed to recover, perhaps because "more river herring may be caught as bycatch in ocean fisheries than the 1 million pounds caught annually in the few remaining fisheries in coastal rivers" (Blankenship). Populations of five North Atlantic deep sea species (roundnose grenadier, onion-eye grenadier, blue hake, spiny eel, and spinytail skate)—all of which are slow-growing fish that can reach 1 meter (3 feet) and live for sixty years—plummeted 87 to 98 percent between 1978 and 1994. Each species is a common bycatch victim; roundnose and onion-eye grenadier, once commercially fished, are now taken almost entirely as bycatch ("Researchers"; D. MacKenzie). Between 1995 and 2004, onion-eye grenadiers declined 93.3 percent; roundnose grenadiers declined "an astonishing 99.6 percent" (D. MacKenzie). We can expect onion-eye grenadiers to further decline, likely as much as 99 to 100 percent "over the next three generations" ("Researchers"). Their average size has already halved because "few fish are getting a chance to mature and breed" (D. MacKenzie). Unless we quit pulling such species from the sea as bycatch, only juvenescence will save them from extinction—though they will remain in a much diminished form.

Shrimp nets are the worst bycatch offenders. The Food and Agriculture Organization (FAO) of the United Nations estimates that 85 percent of any given shrimp haul anywhere in the world is bycatch (Clucas):

Shrimp fishing amounts to only 2 percent of the global wild seafood catch, but is responsible for 30 percent of all the bycatch in the world's fisheries. In some tropical shrimp fisheries, the bycatch is fifteen times the quantity of the shrimp caught. Thailand, the largest source of imported U.S. shrimp, is one of the worst offenders, with a bycatch ratio of 14:1. (Singer and Mason 126)

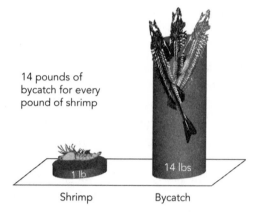

14 pounds of
bycatch for every
pound of shrimp

1 lb   Shrimp

14 lbs   Bycatch

FIGURE 2.3 Shrimp Industry Bycatch

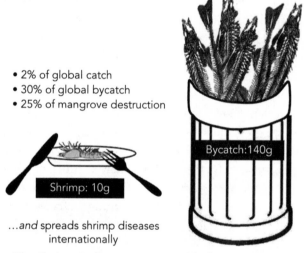

• 2% of global catch
• 30% of global bycatch
• 25% of mangrove destruction

Bycatch:140g

Shrimp: 10g

...*and* spreads shrimp diseases
internationally

FIGURE 2.4 The Shrimp Industry is Responsible for

With every pound of shrimp purchased, consumers pay for 14 pounds
of bycatch that includes threatened and endangered species (Figures 2.3
and 2.4). Shrimp trawling is the leading human cause of mortality among
the earth's endangered sea turtles, destroying more turtles than all other
human causes *combined* (*Decline of the Sea Turtles*).

Increasingly, there are ways for commercial fishers to profit from bycatch,
making matters worse. Some of these otherwise "useless" sea creatures

can now be sold to factory farms—including factory fish farms—and other low-grade fish markets (though threatened and endangered species may not be kept). Consequently, fishers can now profit from these accidental victims, and there is even less incentive to reduce bycatch—especially if bycatch reduction methods might reduce overall catch (Clucas).

Nets

Nets—trawl nets, drift nets, gill nets, seine nets, and so on—are the most common fishing method. Fishing with a net is simple: Sea life swims (or is swept or scooped) into a net and is then pulled from the water. Nets are indiscriminate: They catch endangered species and an abundance of bycatch— anyone who enters their web, including air breathers such as seals, whales, and dolphins (who drown when nets prevent them from surfacing for air).

- Most common fishing method globally.

- **Indiscriminate—**
  - Some 6,000 pinnipeds (seals, sea lions, walruses) and cetaceans (whales, porpoises, dolphins) are killed incidentally every year.

  (clker.com)

  - Some 6 million dolphins have been killed since tuna nets were introduced in the 1950s.

  (public-domain photos.com)

  - Vaquitas, the world's most endangered cetacean (only a few hundred remaining), are disappearing largely because of fishing nets.

  (with permission, Folkens)

s2.6 Methods: Nets (Drift Nets, Gill Nets, Trawling Nets, and Seine Nets)

Some 6,000 cetaceans (whales, porpoises, and dolphins) and pinnipeds (seals, sea lions, and walruses) die in nets, particularly gill nets (Read 163) every year. Most famously, purse seine tuna nets have killed some six million dolphins since these nets were introduced in the late 1950s ("Tuna-Dolphin Issue"). The vaquita, a rare porpoise endemic to the northern part of the Gulf of California, has been decimated by fishing nets. Vaquitas are currently the world's most endangered cetacean—only a few hundred individuals remain. Gill nets are primarily responsible for their decline and near disappearance. Even though reserves have been set aside to protect vaquitas and their essential habitat, these busy swimmers continue to turn up in fishing nets—a minimum of 20–30 each year, an alarming loss for such a small remnant population (Dalton 590, "Vanishing Vaquita" 21, Rodríguez-Quiroz et al. 291).

1. **Cetacean:** aquatic mammals who breathe air through a blowhole, including whales, dolphins, and porpoises.

2. **Pinniped:** carnivorous, flippered mammals, including seals, sea lions, and walruses.

(wpclipart.com)

(clker.com)

s2.7 Key Species Terms

Large sharks, skates, and finfish suffered a 60 percent decline after just five years of industrialized trawling in the Gulf of Thailand (Myers and Worm 281). Industrial midwater trawlers are "the largest fishing vessels on the East Coast of North America, capable of netting 500,000 pounds of sea life in one tow" ("Bycatch and Monitoring").

> The herring industry is becoming increasingly dominated by high-volume industrial ships known as midwater trawlers—which drag massive small-mesh nets behind them, catching everything in their path. The trawlers sometimes work in pairs so they can drag even bigger nets between them. The practice can lead to localized depletion of herring and contribute to the overfishing and stalled recovery of severely depleted populations of cod, hake, haddock, and other fish that live near the ocean floor. ("Under Pressure" 7)

Both American shad and river herring "are thought to be at near-record lows" thanks to herring trawlers off of New England (Blankenship). "[F]ish, birds and marine mammals that feed on herring schools are also vulnerable to accidental capture, injury or death in the trawlers' massive nets" ("Bycatch and Monitoring"). Unfortunately, no one has documented bycatch on herring trawlers in the Atlantic ("Bycatch and Monitoring").

Bottom trawls are pulled along the ocean's floor on large rollers, sweeping up everything in their path and crushing whatever remains. Five metric tons of trawl scraping the ocean's belly leave only churned ground debris in their wake (Deep Sea Conservation Coalition 2). Ecosystems formed over the course of centuries are annihilated in a single pass (Figures 2.5 and 2.6). Though the competition is steep, perhaps the most environmentally damaging netting methods are those of bottom trawls. Clear-cutting does to landscapes what bottom trawls do to ocean floors, but while almost every contemporary environmentalist understands the ill-effects of clear-cutting, few environmentalists can even explain what a bottom trawl is, let alone what it does to sea ecosystems.

Drift nets are also environmentally sinister, and are in many ways analogous to strip mining. Drift nets float free in the ocean, vertically, to a depth of 26 or 34 feet (8–10 m), with weights on the bottom and floats on top. They can be anywhere from 10 to 40 miles (16–64 km) in length, and in the North Pacific there can be as many as 1,800 ships rolling out 28,000 to 35,000 miles (45–56 km) of very efficient drift nets (P. Watson 641): Few squid or fish escape the perilous clutches of their nylon grip.

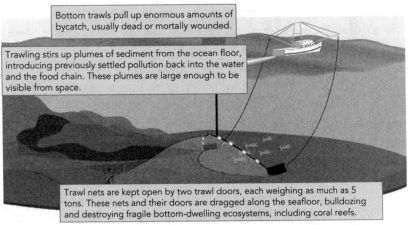

Shrimp trawls have the worst bycatch ratio, but fish trawls can net 500,000 lbs in one tow.

Bottom trawls pull up enormous amounts of bycatch, usually dead or mortally wounded.

Trawling stirs up plumes of sediment from the ocean floor, introducing previously settled pollution back into the water and the food chain. These plumes are large enough to be visible from space.

Trawl nets are kept open by two trawl doors, each weighing as much as 5 tons. These nets and their doors are dragged along the seafloor, bulldozing and destroying fragile bottom-dwelling ecosystems, including coral reefs.

Adapted, included with permission: Stefan Kubicki, Save Our Seas Foundation ("Threat 1")

FIGURE 2.5 Anatomy of a Bottom Trawl

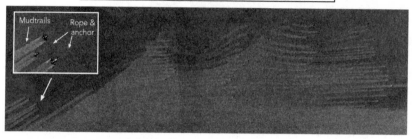

Shrimp trawlers stir up plumes of sediment large enough to be visible from outer space

Included with permission: Van Houtan and Pauly.

FIGURE 2.6 Shrimp Trawlers Stir Up Plumes of Sediment

Death in a drift net is an "agonizing ordeal of strangulation and suffocation," yet "[w]hales and dolphins, seals and sea lions, sea turtles, and sea birds are routinely entangled" (P. Watson 641).

Perhaps even more horrifying, these deadly nets are often lost at sea, becoming "ghost nets" that ensnare and kill sea creatures for as long as their strong nylon fibers last—which is to say, a very long time. Megafleets bearing drift nets "lose an average of six miles of net" every day (P. Watson 643). At present, more than 10,000 miles (16,000 km) of ghost nets are estimated to be floating the seas, luring new victims into their floating glob of decaying matter (P. Watson 643).

Hooks

Droplines and longlines are both common commercial fishing methods; longlines are likely the more popular of the two. About five million longline hooks are dropped in the ocean every day, dangling an estimated one billion (some estimates indicate 10 billion) razor-sharp hooks into the earth's oceans annually ("Long Line Fishing" 1). Japanese fishing interests maintain longlines in every ocean except the circumpolar seas (Myers and Worm 280). Longlines have a long central line, some reaching up to and beyond 50 miles (80 km) "strung with smaller lines of baited hooks, which dangle at spaced intervals" ("Longlining") (Figures 2.7 and 2.8). They are intended to catch large predator fish, such as tuna and swordfish, but are indiscriminate. Blue and white marlin (billfish), "identified as overfished," are frequent victims (J. W. Watson and Kerstetter 9). Longlines hook an estimated 4.4 million birds, turtles, sharks, billfish, and marine mammals incidentally every year (Ovetz 4). Longlines "are considered a critical threat to albatrosses and large petrels" (J. W. Watson

Longlines are held at a predetermined depth by floats and weights, depending on target species. Mainlines run horizontally, dangling shorter, weighted, vertical lines (snoods) at intervals. Each snood has a series of hooks.

FIGURE 2.7 Methods: Hooks/Longlines

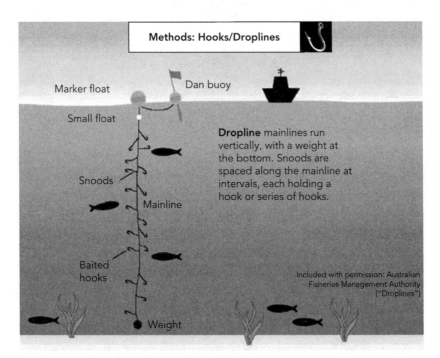

FIGURE 2.8 Methods: Hooks/Droplines

and Kerstetter 9). Nineteen of the world's twenty-one known albatross species are threatened with extinction; to "save the black-footed albatross we need to better regulate Hawaii's longline fisheries" ("Black-Footed Albatross" 6).

Longlines also threaten sea turtles, including loggerhead, leatherback, olive ridley, Kemp's ridley, and green turtles (J. W. Watson and Kerstetter 9). If turtles hooked on longlines do not immediately drown, the physiological stress of being caught and held on a longline affects their "ability to feed, swim, avoid predators, and reproduce" ("Gulf Sea Turtles" 11). Cetaceans also fall victim: Marine mammals most likely to become longline bycatch include pilot whales, Risso's dolphins, and pseudorcas (false killer whales, cousins of the orca that live in and around the Hawaiian Islands, now threatened by longline fishing) (J. W. Watson and Kerstetter 9; "False Killer Whale").

| What About Hooks? | |

- Hooks are common globally—Some 5 million longline hooks ply the oceans every day.
- Hooks target large predators (tuna, swordfish).
- Hooks are indiscriminate—longlines kill 4.4 million sea animals. incidentally every year (seabirds, marine mammals, sharks, billfish, turtles).

(wpclipart.com)

s2.8 What About Hooks?

"Solutions"

Industrial fishing endangers wildlife and destroys sea ecosystems, yet our taste for fish persists. Are there environmentally friendly options for those who wish to continue eating fish?

Aquaculture—Factory Fisheries

Aquaculture—the farming in captivity of fish previously caught only in the wild—seems to promise vast new food production resources that can relieve pressure on overburdened wild fisheries and provide a source of fish as food for the poor. Yet aquaculture raises its own problems. Aquaculture development has resulted in massive changes in land use, polluted neighboring waters with effluent, and spread disease among fish farms. Many studies have warned of the potential risks that escaped farmed fish pose for wild populations. And. . . aquaculture. . . also presents sizable environmental trade-offs by raising demand for wild-caught fish as an ingredient in feed for farmed fish.

—DELGADO

Aquaculture, more commonly referred to as fish farming—and more accurately labeled factory fisheries—is sometimes put forward as a way of satisfying our taste for fish while dodging the environmental problems linked with nets and hooks. "With wild fish production stagnating, growth

Key Fishing Terms

Bottom Trawling
(ESchoolToday.com, "How")

1. **Bycatch**: nontarget sea life incidentally pulled in along with more marketable species.

2. **Trawl:** from 16th century Dutch, a word meaning "to drag"; a fishing method whereby a net is pulled through the water—dragging hooks instead of nets is called trolling ("Trawl").

3. **Aquaculture:** farming aquatic organisms such as fish, crustaceans, mollusks.

s2.9 Key Fishing Terms

Bycatch:
Industrial Fishing is Indiscriminate

(wpclipart.com)

- Hooks and nets catch endangered and protected species.
- Nets catch and drown turtles, dolphins, seals, and whales.
- Fishing lures attract and drown birds.
- Deep-sea fish are pulled up, rupturing blood vessels and air bladders.
- **Bycatch mortality = 90-100%.**

(clker.com)

s2.10 Bycatch: Industrial Fishing is Indiscriminate

in overall fish production has come almost entirely from the global boom in aquaculture, especially in developing countries. Aquaculture now represents more than 30 percent of total food fish production, up from just 7 percent in 1973" (Delgado et al. 7).

But factory fishing also decimates wild fish populations. Factory fisheries generally raise carnivorous fish, such as salmon, and they depend on wild-caught populations, such as "milkfish in the Philippines and Indonesia, tuna in South Australia, shrimp in South Asia and parts of Latin America, and eels in Europe and Japan," to feed carnivorous farmed fish (Naylor et al. 6). "The use of wild fish to feed farmed fish" further depletes flagging fish populations (Jason). And as with factory farming, factory fisheries ultimately burn up more food than they produce because they cycle nutrients through other animals before they reach the mouths of humans. "[F]ish farms consume many more fish than they supply"; it takes two and a half to three pounds of wild fish to produce one eatable pound of farmed salmon (Dawn 158; Pauly and Watson 45).

Stripping the seas of small fish to feed farmed fish creates imbalance, harming ecosystems and disproportionally affecting predators, which depend on smaller species for sustenance. For example, scientists have documented a correlation between anchoveta numbers and the size of sea bird and mammal populations off the coast of Peru (Naylor et al. 7). Similarly, taking capelin, sand eel, and Norway pout from the North Sea, "largely for production of fishmeal, has been linked to declines of other wild fish such as cod and also changes in the distribution, population sizes, and reproductive success of various seal and seabird colonies" (Naylor et al. 7). Almost one-third of wild-caught fish are turned into fishmeal and fish oil, which is "used in feeds for terrestrial livestock and farmed carnivorous fish" (Delgado et al. 4).

Factory fisheries raise sea life in netted-off areas of ocean ecosystems, usurping and transforming key ecological zones. For example, "[h]undreds of thousands of hectares of mangroves and coastal wetlands around the world have been transformed into milkfish and shrimp farms" (Naylor et al. 6). Mangroves are important "nursery habitat for juvenile fish and shellfish," and they protect coastlines from severe storms, help mitigate flooding, and filter and cleanse surrounding waters (Naylor et al. 6). "If the full range of ecological effects associated with mangrove conversion is taken into account, including reduced mollusk productivity in mangroves and losses to seagrass beds and coral reefs, the net yield from these shrimp farms is low—even without considering the use of fish meal in aquaculture feeds for shrimp" (Naylor 6).

Another environmental disaster stemming from factory fisheries is that farmed fish inevitably escape and mate with wild fish, causing "biological contamination."

> Atlantic salmon—the dominant salmon species farmed worldwide—frequently escape from net pens. In some areas of the North Atlantic Ocean, as much as 40 percent of Atlantic salmon caught by fishermen is of farmed origin. In the North Pacific Ocean, more than a quarter million Atlantic salmon have reportedly escaped since the early 1980s, and Atlantic salmon are regularly caught by fishing vessels from Washington to Alaska. Increasing evidence suggests that farm escapees may hybridize with and alter the genetic makeup of wild populations of Atlantic salmon, which are genetically adapted to their natal spawning grounds. This type of genetic pollution could exacerbate the decline in many locally endangered populations of wild Atlantic salmon. In the Pacific Northwest, there is evidence that escaped Atlantic salmon now breed in some streams, perhaps competing for spawning sites with beleaguered wild Pacific salmon. (Naylor et al. 7)

Unnaturally crowded conditions make it more likely that factory-farmed fish carry parasites and diseases, and wild fish are more vulnerable to diseases and parasites than the factory fish that introduce these problems to wild populations (Rosenberg 24). Farmed fish are treated with chemicals and antibiotics (like crowded factory farmed cattle, hogs, and hens), but wild salmon have no such protection. Recent reports indicate

> a direct connection between the aquaculture industry's rapid growth in the Broughton Archipelago off British Columbia and the sharp decline in its wild pink salmon due to an infestation of open-net salmon pens by sea lice. Though older salmon can handle the parasite, young wild salmon migrating through these areas are much more vulnerable. "In the natural system, the youngest salmon are not exposed to sea lice because the adult salmon that carry the parasite are offshore. . . . Fish farms cause a deadly collision between the vulnerable young salmon and sea lice. They are not equipped to survive this, and they don't." ("Salmon Farming" 18, quoting Alexandra Morton)

Transporting factory-farmed sea life also harms individuals and species, threatening ecosystems: Until recently, the white-spot and yellowhead viruses only haunted shrimp farms in Asia, but they have now spread across North, Central, and South America, most likely introduced via a Texas shrimp farm and/or by shipping "contaminated white shrimp larvae throughout the Americas" (Naylor et al. 8).

Key Biology Terms

(clker.com)

1. **Bioaccumulation**: when toxic substances (mercury, uranium, DDT) cannot be processed they accumulate in the body.

2. **Juvenescence**: when a species evolves to reach reproductive maturity earlier in life; in target fish populations, a biological reaction to overfishing.

3. **Biological Contamination**: the result of farmed species escaping and interbreeding with wild species, contaminating wild populations' genetic codes.

s2.11 Key Biology Terms

Additionally, food, waste, chemicals, hormones, and antibiotics from factory farmed fish affect larger local marine ecosystems (Cufone A17). Fish food pellets create "about as many emissions of nitrogen and phosphorus as the poultry industry" (Hawthorne 38). In the waters around factory fisheries, "buildup of food particles and fecal pellets under and around fish pens and cages interferes with nutrient cycling in seabed communities. And when quantities of nitrogen wastes such as ammonia and nitrite are greater than coastal waters can assimilate, water quality can deteriorate to a level that is toxic" (Naylor et al. 8).

Fish farms also bring death to native carnivores, who are attracted to large concentrations of fish. From ospreys to herons, from sharks to seals, from bears to mink, otter, and raccoons—animals that nip into factory fishing profits are generally shot, part of the same predator control program employed on behalf of animal agriculture.

Even if aquaculture were environmentally sound, factory fisheries are morally problematic. Fish farming involves many of the same ethical concerns raised by factory farms—most obviously, suffering from intense confinement. As noted, fish are biologically and socially complex, underscoring the moral imperative for taking such interests into account: If we maintain factory fisheries, we need to develop, implement, and enforce animal welfare standards and laws to protect fish on fish farms ("Ethics & Suffering").

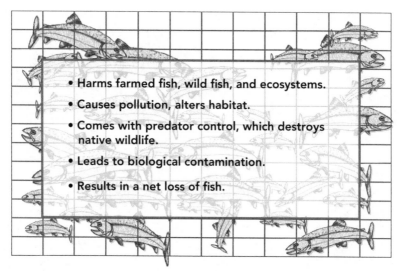

- Harms farmed fish, wild fish, and ecosystems.
- Causes pollution, alters habitat.
- Comes with predator control, which destroys native wildlife.
- Leads to biological contamination.
- Results in a net loss of fish.

S2.12 Aquaculture

Technologies

Increasingly, fishing technologies are designed to reduce bycatch and more specifically, protect endangered species. Net design modifications allow dolphins, turtles, and seals to escape through trapdoors; some are equipped with sound-emitting devices that keep dolphins at a safe distance. Bird-scaring tori lines (streamers) help to keep birds at bay, though critics note that frightening birds in their natural habitat may have unintended, long-term negative side effects (Figure 2.9). Other methods require a change of fishing practices: Longlines can be set deeper in the water, out of sight of hungry birds; fishing at night protects birds who hunt during the day.

Turtle excluder devices (TED) were designed to prevent turtles from being caught by shrimp trawlers. They have greater or lesser success depending on the features of the ocean floor such as debris, vegetation, and rock formations (*Decline of the Sea Turtles*). TEDs involve a series of bars that allow smaller target species to pass into the holding net, while larger turtles bump into bars that lead them to a gate that gives way, offering escape (Figure 2.10) (United States Government). Unfortunately, many fishing boat operators have resisted adopting this technology because TEDs are expensive, and because they are reported to reduce the volume of shrimp caught (*Decline of the Sea Turtles*). In some areas, such as the Gulf of Mexico, TEDs are required by law (Protective Fisheries Technologies).

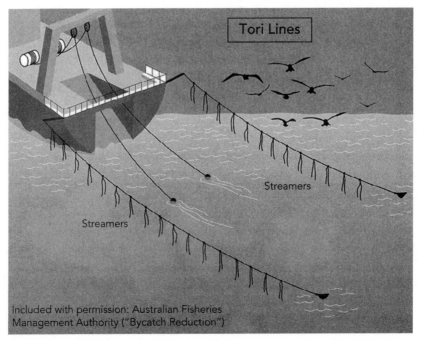

FIGURE 2.9 Tori Lines

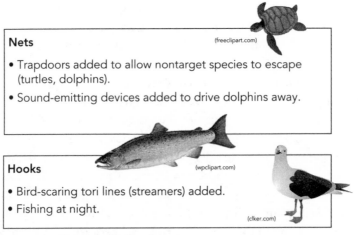

Nets

(freeclipart.com)

- Trapdoors added to allow nontarget species to escape (turtles, dolphins).
- Sound-emitting devices added to drive dolphins away.

Hooks

(wpclipart.com)

- Bird-scaring tori lines (streamers) added.
- Fishing at night.

(clker.com)

S2.13 Protective Fisheries Technology

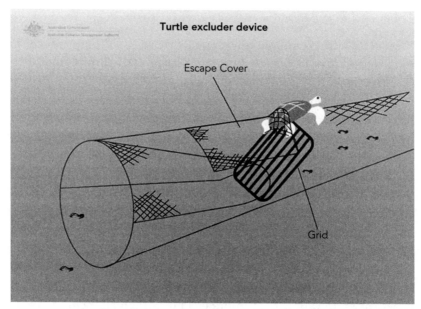

Turtle excluder device

Escape Cover

Grid

FIGURE 2.10 Turtle Excluder Device

Protective technologies are only effective for well-developed fishery management programs—programs that include protected areas, monitoring, and sampling, as well as supervision and enforcement ("Bycatch and Monitoring"). The stripping of the seas is an international problem, and any successful solution must also be international and cooperative. But even with international enforcement, extant technologies address just a tiny fraction of the serious ecological problems that stem from commercial fishing.

> ## Fisheries Technologies
>
> • Are not required.
>
> • Are not enforced. (clipartlord.com)
>
> • Do not address most fishery problems.
>
> • Are effective only within well-developed, international fishery management programs . . . which do not currently exist.

s2.14 Fisheries Technologies

Enhanced Regulations—Too Little, Too Late

It seems a fairly common assumption that we need only reduce the number and size of fishing fleets and lower catch limits, then flagging fish populations will recover. Unfortunately, more comprehensive measures are required to save the ocean's decimated species, especially those that mature slowly: "Harvesting long-lived species on a sustainable basis is much more difficult than we used to believe. Populations don't seem to bounce back just because fishing effort is reduced" (Schiermeier 212).

One of the key problems is fishing methods themselves—indiscriminate methods that tend to destroy habitat. Bottom trawling, for example, is never ecologically sound because it destroys critical habitat. Sponges, coral, and rock formations on the ocean floor provide feeding and breeding grounds for marine life, and protect otherwise vulnerable animals from predators. These fragile habitats do not recover quickly—larger corals grow only 5–25 millimeters (less than an inch) each year, depending on location, while fast-growing corals might manage 20 centimeters (8 inches) per year—and they are already under threat from increased carbon dioxide emissions absorbed by the sea, which affect coral's "chemical building blocks" ("About Coral Reefs"; Feely, Sabine, and Fabry 2): Thirty percent of the earth's corals have perished in the last twenty-five years (Beck 25).

Commercial fishing methods are also indiscriminate, randomly hooking and netting trillions of sea creatures, including innumerable threatened and endangered species. Given the expected ravages of climate change, including the sea's ongoing, increased absorption of carbon dioxide, seas remain under sever, long-term threat from forces other than fishing. We

- **Commercial fishing methods are indiscriminate.**
- **Carbon dioxide** emissions are already damaging vital habitat and, ultimately, sea life.
- **Bottom trawling** destroys vital feeding and breeding grounds:
  - sponge, coral, and rock formations,
  - slow growing coral reefs.

(clipartlord.com, openclipart.org)

s2.15 Technologies and Regulations Do Not Solve Key Problems

must do all that we can to protect seas and sea life. Reducing the number and size of the fishing fleets and lowering catch limits is *not* all that we can do. In a world crowded with humanity, with seas in crisis, fishing is only sensible for those who have no other option, as a means of subsistence—gathering just enough to feed a family, releasing incidental catch immediately.

Marine Reserves

Large marine parks are a viable method of restoring some fish populations and ecosystems. To protect large, slow-growing fish and fragile ecosystems, marine parks must be sustainable, supervised, protected by law enforcement, and completely closed to fishing. Such marine parks can offer fish a safe haven, where even the slowest-growing species can reach sexual maturity. For example, the Chagos Marine Protected Area was established in 2010, in waters under British control in the central Indian Ocean (3,000 km south of India, 3,400 km east of Africa). Chagos is currently the earth's largest marine reserve.

Marine parks are a critical part of any attempt to save the earth's decimated seas. Governments currently pay between $15 and $30 billion a year in fishing subsidies to fishers who have lost their source of income as a result of essential fishing closures and restrictions (Harder 414). It would be far more sensible—ecologically and financially—to pay qualified out-of-work fishers to manage sea parks, engaging their knowledge

Chagos Marine Protected Area

Largest no-take marine reserve in the world

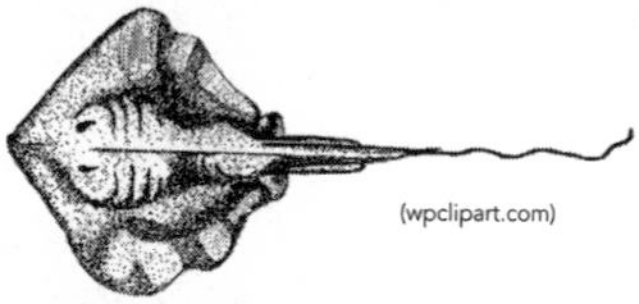

Protects:

(wpclipart.com)

– **640,000 sq km**

– some of the cleanest ocean waters

– world's largest coral atoll (Great Chagos Bank)

– nearly half of the Indian Ocean's most prized reefs, including one of the world's healthiest reefs

FIGURE 2.11 Chagos Marine Protected Area

of marine life and waterscapes to protect and restore ocean ecosystems. Government funds for marine parks can be gathered, in part, by user fees for activities such as snorkeling and kayaking.

Fishers have fished themselves out of a job. It is much more sensible to redirect fishery subsidies to establish and staff marine parks, partly funded by user fees from those wishing to enjoy unfished, protected waters. It makes no sense to subsidize a destructive and dying industry, or to pay out-of-work fishers to do nothing when they could be happily,

- Closed to fishing
- Protections enforced
- Can employ those who previously fished

(clker.com)

s2.16 Marine Reserves

gainfully employed.

Consumer Choices and "Sustainable" Fish Flesh

Demand for and consumption of aquatic flesh has increased steadily in the past century. A number of factors have bolstered consumption, such as human population growth, the birth of aquaculture, misgivings about factory farming, and more knowledge regarding health risks entailed in the consumption of red flesh combined with misinformation about the health value of fish flesh.

Fish consumption continues to grow such that, in 2011, "fish production topped beef production. The gap widened in 2012, with output from fish farming. . . reaching a record 66 million tons, compared with production of beef at 63 million tons" (Larsen and Roney). Growth in fish consumption is an unsettling turn of events given that commercial fishing has *already* devastated ocean ecosystems. A host of oceanic fish are now at risk from overfishing, including red snapper, swordfish, sharks, flounder, sole, Chilean sea bass, blue and white marlins, Pacific salmon, swordfish, smelt, rockfish, herring, grouper, and the critically endangered bluefin tuna ("Seafood Traditions"). A 17-year study across 5 species of deep water fish revealed 87–98 percent population declines (Devine, Baker, and Haedrich 29). Large predatory fish populations (such as tuna, sharks, and marlins) have dropped 90 percent since the introduction of industrial fishing (Myers and Worm 282).

A 40-year study indicates that larger species of sharks have declined precipitously: Sandbar sharks by 87 percent, blacktip sharks by 93 percent, tiger sharks by 97 percent, scalloped hammerheads by 98 percent, and bull, dusky, and smooth hammerhead sharks by 99 percent (or more) (Myers et al. 1848). The longest running shark survey, "conducted annually since 1972" off the coast of North Carolina, "demonstrates sufficiently large declines in great sharks to imply their likely functional elimination" (Myers et al. 1847). The effects of such a loss are far-reaching, and already evident, including a marked increase in the great shark's primary prey species, such as the cownose ray. And as can be expected, the "[e]ffects of this community restructuring have cascaded downward from the cownose ray," which feeds largely on scallop, terminating a "century-long scallop fishery" (Myers et al. 1846).

In response to seas in crisis, some concerned consumers have switched to "sustainable" fish. Unfortunately, there is no dependable source of sustainable fish—seafood labels cannot be trusted. It is a common industry practice, when packaging, to substitute "one seafood species for another," both in the U.S. and abroad, "at levels ranging from 25 to more than 70 percent for commonly swapped species such as red snapper, wild salmon and Atlantic cod" (Warner et al.).

> From 2010 to 2012, Oceana conducted one of the largest seafood fraud investigations in the world to date, collecting more than 1,200 seafood samples from 674 retail outlets in 21 states to determine if they were honestly labeled. DNA testing found that one-third (33 percent) of the 1,215 samples analyzed nationwide were mislabeled. (Warner et al.)

Oceana found that snapper was mislabeled 87 percent of the time, and tuna 59 percent of the time (Warner et al.). With regard to red snapper specifically, only seven of 120 samples nationwide "were actually red snapper"—the remaining 113 samples were "some other species of fish" (Warner et al.). Sushi venues tended to sell mislabeled fish most often (74 percent), "followed by restaurants (38 percent) and grocery stores (18 percent)" (Warner et al.). Deceptive mislabeling foils both health-conscious and environmentally conscious consumers.

> Seafood substitutions included species carrying health advisories (e.g. king mackerel sold as grouper; escolar sold as white tuna), cheaper farmed fish sold as wild (e.g. tilapia sold as red snapper), and overfished, imperiled or vulnerable species sold as more sustainable catch (e.g. Atlantic halibut sold as Pacific halibut). . . .

[T]esting also turned up species not included among the more than 1,700 seafood species the federal government recognizes as sold or

- Aquaculture
- New technologies
- Enhanced regulations
- "Sustainable" fishing

(clker.com)

s2.17 Failed Attempts to Solve Fisheries Problems

likely to be sold in the U.S. (Warner et al.)

It is admirable to seek healthy, sustainable foods. The best way to do so is to avoid animal products, including sea flesh.

Conclusion

Unlike other threatened ecosystems, the impending oceanic crisis is largely unseen and largely unmentioned even among environmentalists—even among ocean specialists: Very few papers published in leading marine ecological journals deal with overfishing (Kochin and Levin 723). Given that we do not see the effects of our voracious and fishy appetites, this is not surprising. "The ocean is our blind spot: a deep, dark, distant, and complex realm covering 70.8 percent of Earth's surface" (Whitty 40). While animal advocates rally on behalf of dolphins and seals, and earth advocates step up to the plate for coral and whales, neither group seems too worried about Chinook salmon or Atlantic halibut—but they should be.

The situation is made worse by a dearth of knowledge. We know less about aquatic species and ecosystems than we do about their terrestrial counterparts. Nonetheless, we tend to have a general sense that the earth's waters, and the quantity of life forms within, are indomitable—as we used to feel about North American forests and wildlife, perhaps most famously, the bison of the Great Plains region. But marine ecosystems are on the verge of ecological collapse, and this is the case *because we choose to eat sea creatures*—lots of them. Human beings are devouring ocean inhabitants at unprecedented rates, destroying fragile marine ecosystems in the process.

No doubt our comparative indifference to Atlantic salmon and onion-eye grenadier stems, at least in part, from the fact that they are out of sight and therefore more easily kept out of mind. When we gaze out to sea, light shines and dances on the surface, and it is easy to believe that all is well—nothing *looks* amiss. Most of us are never personally, directly confronted with the ocean's species depletion—we are simply offered a different type of fish products in our markets and restaurants: When target fish become scarce, fishers simply find a new target, and consumers adjust. Additionally, our geographical separation from sea life blinds us to their individuality and incredible complexity. Finally, and perhaps most importantly, our taste for aquatic flesh flavors our attitude toward fishing, fisheries, and fish.

Whether or not we care, and whether or not we are aware, fish commonly listed on menus are now listed as endangered. Each of us chooses whether or not to consume shrimp, crab, salmon, tuna, prawns, pollock,

- Establish marine parks.
- Do not eat sea life.

(clker.com)

s2.18 Can We Save the Seas?

or other citizens of the seas. In the process, we decide the fate of the earth's oceans. We must choose between plodding sea turtles and wheeling albatross, colorful coral reefs and undulating kelp—or fish for dinner. The only way that you can avoid contributing to the silent collapse of the earth's seas is to choose to eat something else. You decide where you stand every time you sit down to eat.

References

"About Coral Reefs: Factsheet." *International Coral Reef Information Network*. Accessed 29 May 2014. <http://www.qdive.net/pdf/CoralReefFactSheet.pdf>.

"A Look at the Biggest Challenges—and the Way Forward." *Seafood Watch Ocean Issues*. Monterey Bay Aquarium. 2011. Montereybayaquarium.org. Accessed 27 Dec. 2011. <http://www.montereybayaquarium.org/cr/cr_seafoodwatch/issues/>.

"Apparently Starved Gray Whales Found Dead on Calif. Beaches." *Wilmington Morning Star* 19 July 1999.

"Atlantic Bluefin Tuna." *National Geographic*. Accessed 21 May 2013. <http://ani mals.nationalgeographic.com/animals/fish/bluefin-tuna/>.

Balcombe, Jonathan. *Pleasurable Kingdom: Animals and the Nature of Feeling Good*. New York: Macmillan, 2006.

Beck, Michael W. "The Sea Wall Nature Gave You." *Nature Conservancy*, May/June 2013: 24–5.

Bingham, Mike. "Falklands Penguins Starve to Death." Environmental Research Unit. *Falklands.net*. Accessed 26 Dec. 2011. <http://www.falklands.net/ PenguinsStarve.shtml>.

Black, Harvey. "Underwater Suffering: Do Fish Feel Pain? A Study Suggests Fish Consciously Experience Discomfort." *Scientific American* Sept./Oct. 2009. Accessed 22 May 2013. <http://www.scientificamerican.com/article. cfm?id=underwater-suffering-do-fish-feel-pain>.

"Black-Footed Albatross Advances Toward Protection: Seabird threatened by Longline Fishing." *In Brief* Winter 2007–2008: 6.

Blankenship, Karl. "Scientists Suspect Decline of Herring if Result of Bycatch in other Fisheries: Fish Managers Considering Stepped-Up Monitoring and Closing Areas to Trawlers." *Bay Journal* 1 July 2010. Accessed 21 May 2013. <http://www.bayjournal.com/article/scientists_suspect_decline_of_herring_ is_result_of_bycatch_in_other_fisheri>.

"Bluefin Tuna Opens 2013 with Record Auction Price at Tsukiji: 1.78 Million Dollars." *MercoPress* 12 Jan. 2013. Accessed 25 Jan. 2013. <http://en.mercopress. com/2013/01/12/bluefin-tuna-opens-2013-with-record-auction-price-at-tsukiji-1.78-million-dollars>.

Brown, Mark. "Fish Photographed Using Tools to Eat." *Wired.com.* 24 Nov. 2013. Accessed 1. Dec. 2013. <http://www.wired.com/wiredscience/2011/07/fish-tool-use/>.

"Bycatch and Monitoring." *Herring Alliance*. Accessed 23 May 2013. <http://www. pewenvironment.org/uploadedFiles/PEG/Publications/Fact_Sheet/Bycatch_ Monitoring.pdf>.

"Bycatch Reduction Devices." Australian Fisheries Management Authority. Accessed 20 Dec. 2013. <http://www.fish.gov.au/fishing_methods/Pages/bycatch_reduc-tion_devices.aspx>.

Chandroo, K. P., I. J. H. Duncan, and R. D. Moccia. "Can Fish Suffer? Perspectives on Sentience, Pain, Fear and Stress." *Applied Animal Behaviour Science* 86 (2004): 225–50. Accessed 13 Aug. 2009. <http://www.aps.uoguelph.ca/~rmoccia/RDM%20 articles/Fish%20Welfare%20-Chandroo,%20Duncan,%20Moccia%202004.pdf>.

Clark, M. R., et al. *Seamounts, Deep-Sea Corals, and Fisheries: Vulnerability of Deep-Sea Corals to Fishing on Seamounts beyond Areas of National Jurisdiction.*

Cambridge, Eng.: United Nations Environment Programme World Conservation Monitoring Centre, 2006.

Clucas, Ivor. "A Study of the Options for Utilization of Bycatch and Discards from Marine Capture Fisheries." FAO Corporate Document Repository. *FAO Fisheries Circular* 928 (1997). Accessed 24 Jan. 2009. <http://www.fao.org/docrep/w6602e/w6602e00.htm>.

Cufone, Marianne. "Ocean Fish Farms and Public-Resource Privatization." *American Prospect* 19.12 (2008): A17.

Dalton, Rex. "Net Losses Pose Extinction Risk for Porpoise." *Nature* 429.6992 (2004): 590.

Dawn, Karen. *Thanking the Monkey: Rethinking the Way We Treat Animals.* New York: Harper, 2008.

The Decline of the Sea Turtles: Causes and Prevention. Washington, D.C: National AcademyPress, 1990. Accessed 24 Jan. 2009. <http://www.nap.edu/openbook.php?record_id=1536>.

Delgado, Christopher L., Nikolas Wada, Mark W. Rosegrant, Siet Meijer, and Mahfuzuddin Ahmed. *Outlook for Fish to 2020: Meeting Global Demand.* International Food Policy Research Institute, Washington, D.C., WorldFish Center, Penang, Malaysia, 2003. <http://www.hubrural.org/IMG/pdf/ifpri-fish2020-3.pdf>

Deep Sea Conservation Coalition. *The Destructive Power of Deep-Sea Bottom Trawling on the High Seas: Save the High Seas.* DSCC Policy Paper. Deep Sea Conservation Coalition, 24 Sept. 2004. Accessed 10 Jan. 2009. <http://www.savethehighseas.org/publicdocs/DSCC_Bottom_trawling.pdf>.

Devine, Jennifer A., Krista D. Baker, and Richard L. Haedrich. "Deep-Sea Fishes Qualify as Endangered." *Nature* 439.7072 (2006): 29.

Dionys de Leeuw, A. "Contemplating the Interests of Fish: The Angler's Challenge." *Environmental Ethics* 18 (1996): 37–39.

"Droplines." Australian Fisheries Management Authority. Accessed 24 July 2013. <http://www.afma.gov.au/resource-centre/teachers-and-students/about-fishing-methods-and-devices/dropline>.

"Endangered Species in Oregon." *EndangeredSpecie.com.* Accessed 23 Jan. 2013. <http://www.endangeredspecie.com/states/or.htm>.

"ESA Salmon Listings." Northwest Regional Office, NOAA Fisheries. Accessed 8 Feb. 2012. <http://www.nwr.noaa.gov/ESA-Salmon-Listings/>.

"Ethics & Suffering—Sentience and Cognition." www.Fishwelfare.net. Accessed 11 Aug. 11, 2009. <http://www.fishwelfare.net/ethics/sentience.php>.

"False Killer Whale *(Pseudorca crassidens).*" NOAA Fisheries: Office of Protected Resources. Accessed 22 May 2013. <http://www.nmfs.noaa.gov/pr/species/mammals/cetaceans/falsekillerwhale.htm>.

Feely, Richard A., Christopher L. Sabine, and Victoria J. Fabry. "Carbon Dioxide and Our Ocean Legacy." *Pew Charitable Trust Science Brief.* Accessed 23 May. 2013. <http://www.pmel.noaa.gov/pubs/PDF/feel2899/feel2899.pdf>.

"Fishy Recommendations." *Good Medicine* 17.1 (Winter 2008): 23.

Folkens, Pieter. "The Phocoenidae Family: The Porpoise." *Welcome to the Watery World of Whales.* Accessed 13 May 2013. <http://www.whales.org.au/discover/porpoises/index.html>.

"Global Fisheries." *SeaWeb: Leading Voices for a Healthy Ocean. Resources: Ocean Issue Briefs.* Accessed 25 Jan. 2013. <http://www.seaweb.org/resources/brief ings/fishery.php>.

Griesbauer, Laura. "Mercury in the Body and Health Effects." *ProQuest.* Feb. 2007. Accessed 29 May 2014. <http://www.csa.com/discoveryguides/mercury/review4.php>

"Gulf Sea Turtles Get a Breather: Government Orders Review of Long-Line Fishing." *In Brief* Summer 2009: 11.

Harder, Ben. "Cost of Protecting the Ocean." *Science News* 165.26 (2004): 414.

Harrison, John. "Salmon and Steelhead." *Columbia River History Project.* Accessed 23 Jan. 2013. <http://www.nwcouncil.org/history/salmonandsteelhead.asp>.

Harvey, Fiona. "Overfishing Causes Pacific Bluefin Tuna Numbers to Drop 96%: Conservationists also Warned that the Vast Majority Caught were Juveniles and had Never Reproduced." *Guardian* 9 Jan. 2013. Accessed 23 May 2013. <http://www.guardian.co.uk/environment/2013/jan/09/overfishing-pacific-bluefin-tuna>.

Hawthorne, Mark. "Planet in Peril." *VegNews* March-Apr. 2012: 34–41.

"How Does Overfishing Happen?" *EschoolToday.* Accessed 24 Nov 2013. <http://www.eschooltoday.com/overfishing/causes-of-overfishing.html>.

Jason D. "Ecological Impacts of Aquaculture." *Informed Farmers.* 25 Aug. 2011. Accessed 1 Dec. 2013. <http://informedfarmers.com/ecological-impacts-of/>.

Kaufman, Stephen R., and Nathan Braun. *Good News for All Creation: Vegetarianism as Christian Stewardship.* Cleveland: Vegetarian Advocates, 2004.

Kochin, Beth F., and Phillip S. Levin. "Lack of Concern Deepens the Oceans' Problems." *Nature* 424.6950 (2003): 723.

Laland, Kevin N., Culum Brown, and Jens Krause. "Learning in Fishes: From Three-Second Memory to Culture." *Fish and Fisheries* 4.3 (2003): 199–202.

Larsen, Janet and J. Matthew Roney. "Farmed Fish Production Overtakes Beef." *Earth Policy Institute: Plan B Updates.* June 12, 2013. Accessed 30 May 2014. <http://www.earth-policy.org/plan_b_updates/2013/update114>

Lawlor, David. "Preserving our Natural Heritage: The Fish of Earthjustice." *In Brief* Winter 2010/2011: 12–13.

Lehman, Stan. "Hundreds of Dead Penguins Dot Brazil's Beaches" *Huffpost Tech.* 22 July 2010. Accessed 27 Dec. 2011. <http://www.huffingtonpost.com/huff-wires/20100722/lt-brazil-dead-penguins/>.

"Longline Fishing is Killing Sea Turtles." *Sea Turtle Restoration Project.* April 2002. Accessed 2 June 2014. <https://seaturtles.org/wp-content/uploads/2014/06/CCF02062014_00001.pdf>.

"Longlining: Fishing Methods Fact Card." *Seafood Watch.* Accessed 22 May 2013. <http://www.seafoodwatch.org/cr/cr_seafoodwatch/content/media/MBA_SeafoodWatch_Longlining&PurseSeiningFactCards.pdf>.

MacKenzie, Brian R., Henrik Mosegaard, and Andrew A. Rosenberg. "Impending Collapse of Bluefin Tuna in the Northeast Atlantic and Mediterranean." *Conservation Letters* 2 (2009): 25–34.

MacKenzie, Debora. "Deep-Sea Fish Species Decimated in a Generation." *NewScientist: Health.* Accessed 13 Aug. 2009. <http://www.newscientist.com/article/dn8533-deepsea-fish-species-decimated-in-a-generation.html>

"Mad as a Hatter." *Corrosion Doctors.* Accessed 19 May 2013. <http://corrosion-doctors.org/Elements-Toxic/Mercury-mad-hatter.htm#Erethism>.

"Mercury from Cement Kilns to Be Regulated at Last." *In Brief* Summer 2009: 17.

Myers, Ransom A., et al. "Cascading Effects of the Loss of Apex Predatory Sharks from a Coastal Ocean." *Science* 30 March 2007: 1846–50.

Myers, Ransom A., and Boris Worm. "Rapid Worldwide Depletion of Predatory Fish Communities." *Nature* 423.6397 (2003): 280–83.

Naylor, Rosamond L., et al. "Effects of Aquaculture on World Fish Supplies." *Issues in Ecology* 8 (Winter 2001): 1–12.

"Omega-3 Fatty Acids." University of Maryland Medical Center. Accessed 25 Jan. 2013. <http://www.umm.edu/altmed/articles/omega-3-000316.htm>.

"Overfishing: Oceans Are Dying." *Ocean Sentry* 21 Aug. 2009. Accessed 13 May 2013. <http://www.oceansentry.org/en/2557-sobrepesca-muerte-de-los-oceanos.html>.

Ovetz, Robert. *Pillaging the Pacific: Pelagic Longline Fishing Captures and Kills About 4.4 Million Sharks, Billfish, Seabirds, Sea Turtles, and Marine Mammals Each Year in the Pacific Ocean.* Forest Knolls, CA: Sea Turtle Restoration Project, 2004.

"Pacific Salmon, (*Oncorhynchus spp.*)." U.S. Fish and Wildlife Service. Accessed 23 Jan. 2013. <http://www.fws.gov/species/species_accounts/bio_salm.html>.

Pauly, Daniel, and Reg Watson. "Counting the Last Fish." *Scientific American* July 2003: 42–47.

"The Problem with Fishing: Fish and Pain." *Voices for Animals.* PGHMIC: Pittsburgh Independent Media Center. 27 July 2005. Accessed 4 June 2011. <http://pittsburgh.indymedia.org/news/2005/07/19545_comment.php>.

Raloff, Janet. "Empty Nets." *Science News* 167.23 (2005): 360–62.

Read, Andrew J., Phebe Drinker, and Simon Northridge. "Bycatch of Marine Mammals in U.S. and Global Fisheries." *Conservation Biology* 20.1 (Feb. 2006): 163–69.

"Red Snapper FAQs: Biology and Life History." LSU AgCenter: Sea Grant Louisiana. Accessed 14 Sept. 2013. <http://www.seagrantfish.lsu.edu/faqs/redsnapper/biology.htm>.

Roach, John. "Study Calls Into Question Global Quotas on Bluefin Tuna." *National Geographic: National Geographic News.* Aug. 17, 2001. Accessed 29 May 2014. <http://news.nationalgeographic.com/news/2001/08/0817_bluefin2.html>.

Rodriguez, Fernando, Emilio Duran, Juan P. Vargas, Blas Torres, and Cosme Salas. "Performance of Goldfish Trained in Allocentric and Egocentric Maze

Procedures Suggests the Presence of a Cognitive Mapping System in Fishes." *Animal Learning & Behavior* 22.4 (1994): 409–20. Accessed 6 June 2014. <http://download.springer.com/static/pdf/324/art%253A10.3758%252FBF03209160.pdf?auth66=1402250944_9f1cbb4a8a5c9e31bfe5884cd604938a&ext=.pdf>.

Rodríguez-Quiroz, Gerardo, Eugenio Alberto Aragón-Noriega, Miguel A. Cisneros-Mata, and Alfredo Ortega. "Fisheries and Biodiversity in the Upper Gulf of California, Mexico." *Oceanography*. Ed. Marco Marcelli. New York: InTech, 2012. 281–96. Accessed 30 May 2014. <http://www.intechopen.com/books/ocean-ography/fisheries-and-biodiversity-in-the-upper-gulf-of-california-mexico>.

Rollin, Bernard. *Animal Rights and Human Morality*. Amherst, MA: Prometheus, 1981.

Rosenberg, Andrew A. "Aquaculture: The Price of Lice." *Nature* 451.7174 (2008): 23–24.

"Rough Going for the Orange Roughy." *Ocean Portal*. Smithsonian: National Museum of Natural History. Accessed 23 May 2013. <http://ocean.si.edu/ocean-news/rough-going-orange-roughy>.

Sachs, Jeffrey D. "The Promise of the Blue Revolution." *Scientific American* July 2007: 37–38.

"Salmon Farming Threatens Wild Populations." *AWI Quarterly* 57.1 (Winter 2008): 18.

Sample, Ian "Deep Sea Fish Face Extinction." *Guardian* 4 Jan. 2006. 6. Accessed 25 Jan. 2013. <http://www.theguardian.com/science/2006/jan/05/biodiversity.fishing>.

Schiermeier, Quirin. "Europe Dithers as Canada Cuts Cod Fishing." *Nature* 423.6937 (2003): 212.

"Seabirds Needn't Die in Vain." *New Scientist* 195.261 (2007): 6.

"Seafood Traditions at Risk in North America: A RAFT Redlist for Biological Recovery and Cultural Revitalization." Renewing America's Food Traditions (RAFT), Cedar Tree Foundation. Accessed 24 May 2012. <http://www.albc-usa.org/RAFT/images/Resources/SeafoodTraditions.pdf>.

Singer, Peter, and James Mason. *The Way We Eat: Why Our Food Choices Matter*. London: Arrow, 2006.

"SMAST Yellowtail Bycatch Avoidance Program." UMass Dartmouth School for Marine Science and Technology. Accessed 23 May 2013. <http://www.umassd.edu/smast/bycatch/>.

Smith, Rick, and Bruce Lourie. *Slow Death by Rubber Duck: The Secret Danger of Everyday Things*. Berkeley: Counterpoint, 2009.

"The State of World Fisheries and Aquaculture—2006 (SOFIA): Part 1, World Review of Fisheries and Aquaculture." FAO Fisheries and Aquaculture Department. Accessed 22 May 2013. <http://www.fao.org/docrep/009/A0699e/A0699E04.htm>.

"Status Review Report for Atlantic Bluefin Tuna (*Thunnus thynnus*)." National Marine Fisheries Service, National Oceanic and Atmospheric Administration. 20 May 2011. Accessed 25 Jan. 2013. <http://www.nmfs.noaa.gov/stories/2011/05/docs/bft_srr_final.pdf>.

Stenseth, Nils C., and Tristan Rouyer. "Ecology: Destabilized Fish Stocks." *Nature* 452.7189 (2008): 825–26.

Stockstad, Erik. "Global Loss of Biodiversity Harming Ocean Bounty." *Science* 3 Nov. 2006: 745.

"Threat 1: Overfishing." *The Five Threats.* Save Our Seas Foundation. Accessed 28 July 2013. <http://saveourseas.com/threats/overfishing>.

"The Tuna-Dolphin Issue." Southwest Fisheries Science Center: NOAA Fisheries Service: Protected Resources Division. Updated from W. F. Perrin, B. Wursig, and J. G. M. Thewissen, eds., *Encyclopedia of Marine Mammals* (San Diego, CA: Academic Press, 2002), 1269–73. Accessed 22 May 2013. <http://swfsc. noaa.gov/textblock.aspx?Division=PRD&ParentMenuId=228&id=1408#Top>.

"Trawl." *Dictionary.com.* Accessed 6 June 2014. <http://dictionary.reference.com/ browse/trawl?s=t>.

"Turtle Excluder Devices (TEDs)." Australian Fisheries Management Authority. Accessed 25 Nov. 2013. <http://www.afma.gov.au/resource-centre/teachers-and-students/ about-fishing-methods-and-devices/turtle-excluder-devices/>.

"Under Pressure, Government Moves to Protect Herring." *In Brief* Winter 2007–2008: 6–7.

United States Government. National Oceanic and Atmospheric Administration Fisheries Service. Office of Protected Resources. *Turtle Excluder Devices (TEDs).* Accessed 20 June 2009. <http://www.nmfs.noaa.gov/pr/species/turtles/teds. htm>.

Van Houtan, Kyle, and Daniel Pauly. *Kyle Van Houtan and Daniel Pauly's Research on Fishing Trawler Mudtrails.* Accessed 15 Nov. 2013. <http://people.duke. edu/~ksv2/guest/>.

"The Vanishing Vaquita." *AWI Quarterly* 57.4 (Fall 2008): 21.

Viegas, Jennifer. "Video Shows Fish Using Tools." *DNews. Discovery.com.* 29 Sept. 2011. Accessed 1 Dec. 2013. <http://news.discovery.com/animals/ fish-uses-tool-110929.htm>.

Warner, Kimberly, Walker Timme, Beth Lowell, and Michael Hirshfield. "Oceana Study Reveals Seafood Fraud Nationwide." *Oceana* 21 Feb. 2013. Accessed 23 May 2013. <http://oceana.org/en/news-media/publications/reports/oceana-study-reveals-seafood-fraud-nationwide>.

Watson, J. W., and D. W. Kerstetter. "Palegic Longline Fishing Gear: A Brief History and Review of Search Efforts to Improve Selectivity." *Marine Technology Society Journal.* 40.3 (Fall 2006): 6–11. Accessed 22 May 2013. <http://www.sefsc.noaa. gov/turtles/PR_Watson_Kerstetter_2006_MarTechSocJ.pdf>.

Watson, Paul. "Tora! Tora! Tora!" *Environmental Ethics: What Really Matters, What Really Works.* Ed. David Schmidtz and Elizabeth Willott. 2nd ed. New York: Oxford, 2012. 639–43.

Whitty, Julia. "The End of a Myth: Exploring a New Vision for Earth's Greatest Wilderness." *On Earth* Spring 2012: 39–43.

3

Hunting Hype

Hunters attempt to transcend compassion, claiming that acknowledgement of the "truth" of inescapable violence is the highest virtue. It is ironic that this whole position is based on half-truths and distortions.

—LUKE, "A CRITICAL ANALYSIS" 11

WHEN FACED WITH the ecological horrors of animal agriculture, some look to hunting as an escape—as the environmentally friendly way to put meat on the table. This chapter explores the environmental effects of hunting, exposing a handful of myths that help to make this sport appear to be environmentally friendly, animal friendly, socially acceptable—even morally exemplary.

As noted, this book is written specifically for those who have a choice as to what they eat. This book is not a criticism of those who truly have few dietary options (for example, due to affordability or lack of availability).

History: Roosevelt, Pinchot, and Leopold

For millennia men dreamed of acquiring absolute mastery over nature, of converting the cosmos into one immense hunting ground.

—HORKHEIMER AND ADORNO 248

In the United States, wildlife conservation was established *by* hunters *for* hunters *because* of hunters. In the late 19th century, Theodore Roosevelt complained that commercial hunters had decimated wildlife—that a comparatively small population of "market" hunters profited while the nation was stripped of hunter-target species (S. Fox 123). To address these concerns, he founded the Boone and Crockett Club (BCC) in 1897, with the following mission: "[T]o promote the conservation and management of wildlife, especially big game, and its habitat, to preserve and encourage hunting and to maintain the highest ethical standards of fair chase and sportsmanship in North America" ("About the B & C Club"). "Conservation" is a utilitarian, human-centered term promoting the protection of wildlife and wilderness *for human use.* Accordingly, the BCC promoted laws protecting "every citizen's freedom to hunt and fish," and established wildlife as "owned by the people and managed in trust for the people by government agencies" ("About the B & C Club"). As a result of the BCC, the *U.S. government was placed in charge of managing wildlife on behalf of hunters.*

The agenda and actions of the BCC (in conjunction with Roosevelt, a BCC member wielding the presidential pen), led to the creation of the U.S. National Forest System and the National Wildlife Refuge System. Roosevelt assigned his friend, Gifford Pinchot, to the nation's freshly established National Forest Commission—Pinchot was a BCC member and an avid hunter. His family had earned a fortune logging, and, not surprisingly, he viewed the natural world as a storehouse for "resources" that should be "managed" to offer the greatest good to the greatest number over the long run. Though he did not explicitly say so, Pinchot was only concerned about the greatest good for human beings: He viewed forests as means for *human* ends, and felt that U.S. forests ought to remain open for development. Other members of the National Forest Commission objected to Pinchot's utilitarian philosophy, preferring to promote public lands as "locked reserves," not to be exploited ("Gifford Pinchot"). But in 1905, President Roosevelt chose Pinchot as the first head of the newly established Forest Service, sealing the fate of U.S. forests—"resources" to be "conserved" for human ends.

Initially, there were no hunting regulations, and many hunted as if wildlife were a fount that would never run dry, "wiping out some species and reducing others to a pitiful remnant of their original numbers" ("Federal Aid Division"). Roosevelt and his cohorts were determined to preserve hunting (as opposed to ecosystems or individuals), and they developed policies to protect hunter-target species. As part of his effort to preserve and protect hunting interests, Roosevelt founded the National Wildlife Refuge System (NWRS), with a network of 55 "game" reserves. The National Wildlife Refuge System specifically protects target species for hunters.

President Roosevelt felt that hunting was critical to all young men. He maintained that hunting had a civilizing effect, and therefore advocated hunting as an outlet for virile male impulses. His model of "manhood and good sportsmanship" focused on the gentleman hunter, who killed wildlife only according to prescribed rules, showing restraint by honoring the rule of "fair chase." He did not advocate hunting as a means of subsistence or sustenance, but as a moderating influence for man's deadly urges "through the development of the rules of good sportsmanship" (Kheel, *Nature Ethics* 70). Roosevelt also believed hunting to be the natural heritage of all citizens.

While Roosevelt had a decisive and lasting effect on government wildlife conservation agencies and programs in the U.S., Aldo Leopold (author of *A Sand County Almanac*, 1949) is most often touted as the father of conservation. Leopold held much in common with Roosevelt (and Pinchot): He was also an early BCC member who viewed wilderness as "hunting grounds"—in fact, he wrote an article in 1925 which he titled, *"A Plea for Wilderness Hunting Grounds"* ("Excerpts"). He viewed hunting as an "expression of love for the natural world" and a primary reason to preserve wilderness ("A Leopold Biography"). Hunting was Leopold's "first and primary motivator to work in game protection and management," and he remained a "committed hunter until the day he died" ("A Leopold Biography").

Also like Roosevelt, Leopold endorsed hunting as important for the proper development of young men—a means by which they honed "self-reliance, hardihood, woodcraft, and marksmanship—pitting [themselves] against the elements" ("A Leopold Biography"). He viewed this struggle with nature as critical for the proper development of "man," and wildlife as expendable toward this end. Leopold defended hunting as *the* channel through which men developed an intimate connection with and appreciation for nature. In fact, he felt so strongly about the value of hunting with regard to the honing of outdoor skills that his writing reveals a lack of respect for those who do not hunt: "The deer hunter habitually watches the next bend; the duck hunter watches the skyline; the bird hunter watches the dog; the non-hunter does not watch" (Leopold, *A Sand County Almanac* 224).

Most famously, Leopold developed and expounded "the land ethic," which enlarged the "boundaries of the community to include soils, waters, plants, and animals, or collectively: the land" (Leopold, *A Sand County Almanac* 204). According to Leopold's land ethic, a "thing is right when it tends to preserve the integrity, stability, and beauty of the biotic community. It is wrong when it tends otherwise" (Leopold, *A Sand County*

Almanac 224–25). Despite this, Leopold's view of wildlife and wilderness was fundamentally exploitive. In his utilitarian assessment (as with Roosevelt and Pinchot), wildlands and wildlife serve their rightful purpose through the hunt—they were God-given "resources," every hunter's heritage, rightly exploited as a means to "man's" ends.

Roosevelt, Pinchot, and Leopold loved to hunt and held utilitarian views of wildlife and wilderness, and they worked diligently to preserve and protect their favored pastime, "and the avenue to their manhood" (Kheel, *Nature Ethics* 96). Wildlife was "conserved not because of the animals' right to life, but rather because of 'man's' inalienable right to hunt and kill" (Kheel, *Nature Ethics* 120). Still today, hunting in the U.S. is regulated (and encouraged) by the U.S. government, based on foundational U.S. wildlife legislation established by these three men, whose hunter-friendly, utilitarian views of wildlife and wilderness still hold sway.

Government Misappropriations and Deceit

The hunter, seeing there would soon be nothing left to kill, seized upon the new-fangled idea of "conservation." . . . The idea of wildlife "management"—for man, of course—was born. Animals were to be "harvested. They were to be a "crop"—like corn.

—CLEVELAND AMORY, *Man Kind? Our Incredible War on Wildlife*

Wildlife policies in the U.S. are fundamentally utilitarian, shaped by and for hunters. U.S. federal and state government conservation programs, managed by Fish and Wildlife Services (FWS), continue to manipulate wildlife as a "resource" for hunters. This ongoing alliance between government wildlife agencies and hunters tends to be supported by environmentalists because so many people assume that hunting is ecologically friendly—even essential—but hunting is most certainly not environmentally friendly.

U.S. Government Misappropriations

The U.S. government misappropriates funds in order to cater to hunters, damaging ecosystems and betraying public trust in the process. The most flagrant example of this misappropriation and betrayed trust is Wildlife Services wildlife "management."

Wildlife Services—Ecosystem Manipulation on Behalf of Hunters/Ranchers

[I]t hardly needs to be argued that, with a game supply as low as it is, a reduction in the predatory animal population is bound to help the situation. . . . Whatever may have been the value of the work accomplished by bounty systems, poisoning, and trapping, individual or governmental, the fact remains that varmints continue to thrive and their reduction can be accomplished only by means of a practical, vigorous, and comprehensive plan of action.

—LEOPOLD, "Varmint Question"

Human beings traditionally have enslaved those animals they deem worthwhile and set out to eliminate the rest.

—ROBERTSON 93

Leopold was among those who believed that "there could not be too much horned game, and that the extirpation of predators was a reasonable price to pay for better big game hunting" (Leopold, "Review" 322). On behalf of such hunter interests, wildlife agencies started a "frenzied campaign to rid the North American continent of its natural predators" (Robertson 68). In 1913 the U.S. government passed the Animal Damage Control Act, authorizing the U.S. Secretary of Agriculture to suppress, eradicate, and control any non-human animal deemed injurious to human interests, namely the interests of ranchers and hunters (Di Silverstro in C. Fox, "Carnivore Management" 21). U.S. Animal Damage Control has since been given a less descriptive title—U.S. Department of Agriculture's Wildlife Services Program. This deceptive new title suggests serving or helping—serving or helping whom? Certainly not wildlife, but rather ranchers and hunters!

Wildlife Services kills wildlife on behalf of powerful human interest groups, most notably hunters and ranchers. For example, commercial outfitters gained the backing of politicians in their quest to kill 60 percent of the wolf population in Idaho. Their reason? "They want elk herds to grow so that the outfitters' clients—recreational hunters—can kill elk" (Birdseye 12). In the words of a disillusioned local, gunners and politicians "are trying to use a wilderness area, where natural processes are supposed to take place, to create a rugged elk farm" (Birdseye 12). But this wilderness, "the nation's largest forested wilderness" (2.4 million acres), is critical not because hunters can easily kill elk, but because it permits "predators such as wolves to exist without conflicting with livestock"—without conflicting with the interests of Big Ag (Birdseye 13). As it turns out, Big Guns are no less problematic than Big Ag; hunters hope to "seize the wilderness for their own purposes" (Birdseye 13).

No doubt much to the disappointment of hunters such as those in Idaho, Wildlife Services no longer seeks to eradicate species, but rather to keep the numbers of certain "undesirable" or "unprofitable" species in check, while bolstering numbers of hunter target species. Wildlife Services thereby employs hunters and trappers to kill a certain number of individuals from predator populations with "poisons, steel-jaw leghold traps, strangulation neck snares, denning (the killing of coyote pups in their dens), hounding, shooting, and aerial gunning" (C. Fox, "Predators in Peril"). Without the backing of scientific studies, using indescriminate methods, and without prioritizing nonlethal methods (or any manner of humane methods), Wildlife Services kills roughly 1.5 million animals each year ("Petition to Reform" 25). Wildlife Services thereby causes tremendous suffering, and targets and kills thousands of ecologically critical animals, such as coyotes, foxes, bobcats, otters, wolves, black bears, and mountain lions. Wildlife Services exterminates "more than 2.4 million animals each year, including more than 120,000 native carnivores at an annual cost to taxpayers of over $115 million" (C. Fox, "Carnivore Management" 21; C. Fox, "Predators in Peril"). Such "ecologically short-sighted conduct" has inevitably altered ecosystems and extirpated species—some of which have been reintroduced at taxpayer expense, including swift foxes and wolves (Robertson 12). If reintroduced species thrive, wildlife agencies again subject them to a hunting and/or trapping

- poison
- steel-jaw traps
- snares
- conibear traps
- denning (kills pups in dens)
- hounding
- hunting
- aerial gunning

(Clker.com)

s3.1 Wildlife Services: Methods

season, bolstering the profits and pleasures of trappers without compensating the vast majority of taxpaying citizens.

Maintaining "big game" was Leopold's "first and primary motivator" for wildlife "management," and this is still true for federal and state wildlife agencies ("A Leopold Biography"). One major problem with this predator extermination plan (aimed at bolstering big game) is that the birthrates of many species vary *directly* in response to external pressures:

> [W]hen populations are low with respect to the maximum number of individuals an environment can support . . . birth rates (the number of live births per female per year) have a tendency to be high. When a population is at or near the maximum number the environment can support, birth rates are low and death rates (the number of animals in the population dying per year) are high. (Yarrow)

Killing coyotes, for example, ultimately causes them to bear more young than they otherwise would. Consequently, "culling" is ineffective because it guarantees greater resurgence.

- Birthrates often vary *directly* in response to external pressures.
- "Culling" (killing) thereby causes a resurgence.

(Clipartlord.com)

- **In contrast: Immunocontraceptives reduce populations 72-78%.**

s3.2 Animal Damage Control: Self-Perpetuating

If Wildlife Services wishes to manipulate wildlife populations, there are more effective methods than hunting or trapping. For example, immunocontraceptive techniques reduce a variety of populations by 72–86 percent ("Caught in the Crosshairs" 23). Why have state wildlife agencies terminated successful immunocontraception programs? And why were these successful programs smeared with "a campaign of obstruction, misinformation, and outright deception to derail the development, expansion and field validation of immunocontraception" ("Caught in the Crosshairs" 24)?

Hunting was Leopold's "first and primary motivator to work in game protection and management," and to this day "hunters form a large contingent of those who administer wildlife management regulations" ("A Leopold Biography"; Luke, *Brutal* 108). Hunter influence therefore remains strong in federal and state wildlife agencies, and from their perspective taxpayer dollars are well spent supporting "wildlife control"—gunners who pop predators and trappers who ensnare wildlife, working to enhance hunter-target species. Such jobs are choice in the eyes of those who enjoy shooting, trapping, and snaring, while immunocontraception is anathema. Those at the helm are fundamentally uninterested in alternative methods of wildlife management, so Wildlife Services continues to hire gunners and trappers to shoot, trap, snare, and poison specific species, even in the absence of studies assessing the environmental impact of such wildlife manipulation. As long as U.S. Department of Agriculture's Wildlife Services Program exists, and as long as this agency views certain species as a "threat to livestock and a competitor for game species," these animals will be shot, trapped, or poisoned—at taxpayer expense (C. Fox, "Coy Coyote"). For the sake of using words with a measure of integrity, Wildlife Services "wildlife management" ought to be called "wildlife manipulation," and will be for the remainder of this book.

Authorized the U.S. Secretary of Agriculture to suppress, eradicate, and control any nonhuman animal deemed injurious to human interests, namely the interests of ranchers and hunters.

- Established in 1913
- 2.4 million animals exterminated annually
  - > 120,000 are native carnivores,
    - including nontarget species—
      - – endangered species
      - – raptors
      - – pets
- annual taxpayer cost > $115 million.

(WPClipart)

s3.3 Wildlife Services: Animal Damage Control Act

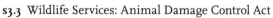

Funding: Taxpayer Burdens, Hunter Benefits

The 100 million animals killed by hunters each year are killed not for wildlife management science but to support a $22 billion firearms industry.

—MOSS

Since the late 19th century, hunters concerned about the future of wildlife and the outdoor tradition have made countless contributions to the conservation of the nation's wildlife resources. . . . Hunting organizations contribute millions of dollars and countless hours of labor to various conservation causes each year.

—U.S. FWS WEBSITE, "Hunting.

Many people assert that conservation exists—and wildlife remains—thanks to hunters. While there is some truth to this assertion historically, it is also true that government wildlife agencies were essential *because* of hunters.

Wildlife conservation began *because of and for* hunters: Were there no hunters in Roosevelt's day, no conservation would have been necessary. Conservation was initiated to protect target animals from hunters—to conserve wildlife for the pleasures of future hunts. Preservation (preserving wilderness for its own sake), as opposed to conservation (conserving "resources" for human use), would eventually have been necessary due to human population growth and expansion, but such preservation would have been designed to protect ecosystems rather than protect hunter interests.

Despite the fact that Wildlife Services caters to hunters, hunters do not foot the majority of the bill for wildlife manipulation. A brief history of the Pittman-Robertson Act explains why this is the case.

Despite Roosevelt's assertion that wildlife declined because of commercial hunters, sport hunters also threatened wildlife populations, and further protections were needed by 1937 to save hunter-target species, so Congress passed the Federal Aid in Wildlife Restoration Act, commonly called the Pittman-Robertson Act (P-R). This Act added an 11 percent excise tax on rifle, shotgun, and ammunition sales, which was earmarked for use by state wildlife agencies to manipulate wildlife in the hope of increasing hunter-target species. States were required to match "grant funds with at least one dollar for every three federal dollars received" ("Pittman-Robertson" 3). As a result, states that sold more guns received more Pittman-Robertson monies, and had more P-R funds available for wildlife manipulation.

The P-R Act effectively tied the interests of state wildlife agencies to the interests of firearms industries. As hunting became more and more of a pastime, and less and less necessary for sustenance, hunter-generated revenues declined. In 1970, the Pitman-Robertson Act was amended, adding "provisions for the deposit of the 10 percent tax on pistols and revolvers" ("Digest"). This amendment further tied government wildlife agencies to the firearms industry, but radically altered *who* pays the P-R tax.

- Early 20th Century: Government regulation necessary to protect hunter target species.

- 1937: Pittman-Robertson (PR) Act—11% excise tax placed on rifles, shotguns, and ammunition to fund government management of hunter target species.
- Mid-20th Century: Hunting declined, leading to a decline in management funds.
- 1970: PR Act amended—10% excise tax placed on pistols and revolvers.

(clker.com)

U.S. excise tax on handguns (for personal protection) is unjustly used to "manage" wildlife on behalf of hunters.

s3.4 History of Funding: Wildlife "Management"

Today less than 5 percent of U.S. citizens hunt, but there are approximately 90 guns for every 100 citizens (Moss; "Hunters"; MacInnes). The most common reason for keeping a gun in the U.S. is now personal protection—usually a pistol or revolver. The second most common reason for owning a gun is target practice (Carroll). Millions of women in New York, Boston, Los Angeles, Chicago, Detroit, San Francisco, and Philadelphia—who have never hunted and never will hunt—purchased handguns and paid the P-R tax. Some of these gun owners (like my sister) vehemently detest hunting, but nonetheless supported the P-R tax with a handgun purchase. Given that less than 5 percent of U.S. citizens hunt, and given that FWS P-R tax now draws largely from those who carry handguns for personal protection, government wildlife agencies *misappropriate* P-R funds when these funds are used to manipulate wildlife and ecosystem on behalf of hunters.

Why do government agencies continue to use nonhunter funds to manipulate wildlife on behalf of hunters? The simple answer seems to be public ignorance. Few citizens know the history of the P-R excise tax, or the damaging ecological effects of federal and state government wildlife

• 10% excise tax on handguns (purchased for personal protection) pays for hunter education and recruitment.

• For every 100 U.S. citizens there are 90 guns (and only 5 hunters).

Nonhunters (>95%) must pay for wildlife management on behalf of hunters (< 5%) to keep the current system running.   (clker.com, wpclipart.com)

s3.5 Who Pays for Wildlife Management?

manipulations on behalf of hunters. In fact, most people—including those working in U.S. wildlife agencies—continue to believe that hunting licenses fund essential wildlife and wildlands protection programs, and that these programs only exist thanks to hunter funding. And as the quote at the beginning of this section demonstrates, U.S. FWS continues to propagate this misconception.

In truth, there are now too few hunters and trappers to fund contemporary U.S. wildlife manipulation programs. Hunting has "been on a downward slide" since about 1970 (Robertson 152). Hunters have "withered and shrunken" to "a wee fraction" of their former numbers (Robertson 154, 153). The average age of a hunter is now "about 46 and keeps inching up" (Heffernan 25). The purchase of hunting and fishing licenses has either remained steady or (more often) declined since 1990. According to Fish and Wildlife, the number of fishers and hunters fell between 2001 and 2006, from 37.8 million in 2001 to 33.9 million in 2006—despite an overall U.S. population increase of 15 million (2006 4; "2006 National Survey"). Simultaneously, as might be expected, their economic investment dropped by three billion. "Literally, figuratively and statistically, hunting is a dying sport" (Robertson 153). Though "a shrill and voluble" minority, hunters constitute a tiny and declining fraction of the U.S. population.

Instead of rethinking their alliance with a dying and miniscule population, FWS (which employs many hunters) requested tax monies to recruit a fresh supply of hunters (Heffernan 25). A willing Congress authorized states to spend *up to half* of P-R revenues on target ranges and hunter "education" programs designed to encourage newcomers to give hunting a shot ("Digest"; "Pittman-Robertson" 1, 3, 5).

Many hunter recruitment programs target women and children. For example, Maine allows "pre-season bow hunting for children ages 10 to 16"; Alabama opens deer hunting season "two days early for children under 16," giving young hunters "a better crack at 'bagging' one" (Robertson 155). Illinois and Colorado offer hunting lessons for single mothers; the NRA supports "Women on Target," while various states foster Becoming an Outdoors Woman (BOW) (Robertson 155). The director of the Alabama BOW program explains why prohunter programs target women: "If you teach a man to hunt, he goes hunting. If you teach a woman to hunt, the entire family goes hunting. Bringing women into the sport helps us reach so many more people. Children who are brought up hunting will likely pursue the activity for their entire lives" (Roney). Recruitment programs have proven somewhat effective: There has been growth "among the youngest female hunters . . . those aged 6 to 15" (Roney). What does each new recruit cost taxpaying handgun owners—and why exactly are they interested in bolstering hunter numbers? When did those purchasing handguns agree to pay for programs that encourage little girls to wield big weapons with intent to kill?

Roosevelt and his cohorts established a system whereby "those who use the resource . . . pay for its care and maintenance" ("About the B & C Club"). This policy was reflected in the initial P-R tax, which required hunters to support conservation—because hunters had decimated hunter-target species and had a vested interest in restoring and conserving these species ("About the B & C Club"). The P-R tax was appropriately placed on guns because those who purchased guns generally acquired them for hunting. But in the 21st century things look quite different: U.S. citizens likely have little interest in manipulation of wildlife or ecosystems by federal or state agencies on behalf of hunters. It is equally unlikely that the majority of U.S. gun owners want excise taxes from gun purchases to be spent training mothers and children to kill for sport, especially in light of the unnerving proliferation of high-profile gun violence incidences in schools.

Perhaps most fundamentally, rather than change their course in light of a significant decline in hunting, government agencies have invested yet *more* money into hunters and hunting. Would the U.S. public, if informed, choose to continue supporting government wildlife agency programs that recruit hunters—targeting women and children—with the hidden

agenda of funding outdated, environmentally damaging wildlife manipulation programs? *Contemporary government agency hunter-recruitment*
strategies would likely seem unscrupulous even to the founders of U.S. Wildlife
Services programs.

Luckily, an alternative source of funding is available. While hunters and hunting have declined steadily over the last fifty years, wildlife watchers swelled one million *each year* between 1996 and 2001
(Robertson 153). This trend has continued: "The ratio of non-hunting
outdoor enthusiasts to hunters grew more than 26 percent in the last
ten years"—there are now six times as many nonhunting wildlife
enthusiasts as there are hunters, despite Fish and Wildlife's investment in a plethora of hunter-recruitment programs (Robertson 153). In
2008, even FWS was compelled to admit that "the participation rates
and economic impact of hunting and fishing now trail those of wildlife watching" (2006 5; Reed). Their 2011 survey (conducted every five
years) analyzing and summarizing "wildlife associated recreation"
discovered that the weight of outdoor enthusiasts does not rest with
hunters: "[O]ver 90 million U.S. residents sixteen years old and older
participated in wildlife-related recreation," but only 13.7 million hunted,
while "71.8 million participated in at least one type of wildlife-watching
activity including observing, feeding, or photographing fish and
other wildlife in the United States" ("2011 National Survey"). Hunters
invested a mere 33.7 billion in their sport, while a robust population
of wildlife watchers injected a hefty 54.9 billion dollars into the econ-
omy ("2011 National Survey"). Wyoming exemplifies the trends of the
nation. Though not known for liberal politics (especially with regard
to guns), twice as many people now choose to watch Wyoming wildlife as choose to hunt Wyoming wildlife: While hunters in 2006 spent
only $138.5 million in Wyoming, wildlife watchers spent a whopping
$392.4 million ("Helping Them All" 1; 2006 96, 102).

Unlike citizens in Roosevelt's time, non-hunters buy everything from
"bird food to binoculars, from special footwear to camera equipment,"
spending large sums of money simply "to enjoy wildlife" ("Federal Aid
Division"). If a 10 percent excise tax on tents, cameras, hiking boots, climb-
ing gear, sleeping bags, binoculars, backpacks, life preservers, skis, and
canoes funded wildlife protection, then federal and state wildlife agencies would shift from conservation on behalf of hunters and hunting to
genuine preservation on behalf of wildlife and wildlands. Contemporary
outdoor enthusiasts prefer to visit intact ecosystems, to watch wildlife on

the hoof—they want to see deer and ducks alive on public lands, not dead in the back of a hunter's truck.

Instead, an excise tax on gun-purchases continues to fund FWS wildlife programs, and they continue to cater to hunters. Consider the five target concerns for "wildlife restoration" programs listed in the 2013 Pittman Robertson "Program Overview" of "Budget Justifications":

- Restoration, conservation, management, and enhancement of wild bird and mammal populations;
- Acquiring and managing wildlife habitats;
- Providing public use of wildlife resources;
- Educating hunters on conservation ethics and safety; and
- Constructing, operating, and managing recreational shooting and archery ranges ("Pittman-Robertson" 3).

First, we might wonder why certain species need to be "restored" in the first place. Second, we might notice that each of these goals is designed to attract and benefit hunters, as well as cover the tracks of the hunter's environmental footprint. We might also wonder, other than hunters and trappers, who has a vested interest in "using" wildlife resources? Outside of hunting and trapping what would that mean?

What would wildlife agency goals look like if they did *not* focus on hunter interests, instead reflecting the interests of the majority of citizens? Perhaps they would look something like this:

- *Preserve* wildlife and habitat.
- Acquire and *preserve* additional habitat.
- Maintain *public access to protected habitats*, and
- Educate *visitors* on the importance of wilderness and wildlife *protection and preservation*.

Given the decisive shift in public interest in the last decades, it is no longer just or sensible (if, indeed, it ever was) for wildlife agencies to cater specifically to hunters. From hunting-license fees to ammunition, from government-funded hunter-recruitment to wildlife manipulation programs, from ATVs to high-powered rifles, hunting in the U.S. is big bucks—greenback bucks. Nonetheless, FWS's ongoing alignment with hunting interests and the firearms industry constitutes a breach of public trust and a gross misappropriation of public funds.

"Public" Lands

The Mission of the National Wildlife Refuge System is to administer a national network of lands and waters for the conservation, management, and where appropriate, restoration of the fish, wildlife, and plant resources and their habitats . . . for the benefit of present and future generations of Americans.

— "MISSION," National Wildlife Refuge System

At the turn of the 20th century, white-tailed deer sightings were so rare they were reported in daily newspapers. Now, thanks in part to hunter conservation, deer populations are thriving, and national wildlife refuges are great places to hunt.

— "HUNTING DEER," FWS website

Though the government still insists that the "majority of public lands in the United States are held in trust for the American people by the federal government," in light of a drastic decline in hunters, hunter-friendly policies can no longer be considered management "in trust for the American people" (if ever they could) ("Public Land"). Nonetheless, hundreds of millions of acres of public lands are managed specifically for hunters, most notably the nations 556 Wildlife Refuges ("Statistical").

By definition, a "refuge"

- offers "shelter or protection from danger or distress,"
- is "a place that provides shelter or protection,"
- offers "recourse in difficulty" ("Refuge").

- "shelter or protection from danger or distress."

- "a place that provides shelter or protection."

- "recourse in difficulty."

(clker.com)

s3.6 Refuge Defined

Merriam-Webster offers the following example for the proper use of "refuge": "[H]unting is strictly forbidden in the wildlife *refuge*" ("Refuge"). Indeed, given the definition of "refuge," most people familiar with English would expect a "wildlife refuge" to be a place where wild animals are sheltered and protected, but most U.S. National Wildlife Refuges are decidedly dangerous places for wildlife.

Hunting is listed first among the agency's "[w]ildlife-dependent uses," followed by fishing ("Guiding Principles"). The home page of United States National Wildlife Refuge System (NWRS) website bears the heading "Your Guide to Hunting on National Wildlife Refuges," and includes a search engine to help people "Find the Perfect Hunt" ("Your Guide"). The NWRS website also lists the agency's "Guiding Principles," including "wildlife management," defined as "active *manipulation* of habitats and populations, [as] necessary to achieve Refuge System and U.S. Fish and Wildlife Service missions" ("Guiding Principles," italics added). The Wildlife Refuge System includes lands labeled as "Waterfowl *Production* Areas" ("Statistical," italics added). U.S. wildlife refuges were established to "create well-stocked hunting grounds," as a "breeding ground" for "game for sport hunting" (Kheel, *Nature Ethics* 112). But manipulating public lands on behalf of hunters, who constitute less than 5 percent of the overall population, is in gross violation of the stated task of preserving public lands and their wild residents on behalf of "present and future generations of Americans" ("Mission"). In fact, ongoing government wildlife manipulations are disadvantageous for ecosystems *and* wildlife, and are therefore decidedly not in the interests of present or future Americans.

Public lands are more generally hunter-friendly at the expense of the larger public. Opening public lands to hunters puts all visitors at risk—during hunting seasons, those visiting public lands are expected to acquire and wear fluorescent orange. In fact, hunters put other citizens at risk even when those other citizens are on their own property, or when they are walking on a public road. In Maine, Karen Wood was shot dead in her backyard by a hunter who mistook her for a deer. Her killer was acquitted (Gross). In McLeansville, North Carolina, a hunter shot a man and his 13-year-old son as they walked along a local road ("Police"). Law enforcement attributed these slayings to poor light and to the hunter's "failure to confirm his target," but the police found no grounds to arrest the shooter ("Police"). In Reamstown, Pennsylvania, shotgun pellets were fired into a parking lot and struck a woman in the eye, but a game warden noted that there would be "[n]o jail time—not for something like this," because it was "an accident" ("Woman Shot"). While citizens who wish to drive a car must repeatedly pass a mandatory vision test, there is no such requirement for hunters. Why not? And why does reckless driving land us in jail, but not reckless hunting?

In addition to fearing for their lives, citizens visiting public lands during hunting season know that they will not have much of a chance to view wildlife, especially terrified, hunter-target species. And thanks to wildlife manipulation, the public *never* sees natural ecosystems—ecosystems unaffected by hunters and fishers—on public lands.

There are several possibilities for justly managing public lands on behalf of the American people. For example, hunting can be restricted to a percentage of our public lands so as to reflect the percentage of hunters—less than 5 percent. This would leave 95 percent of public lands open for general use year-round for the non-hunting majority (95 percent). Alternatively, hunting could be restricted to a population-proportional percentage of time—5 percent, allowing the vast majority (the 95 percent) to feel safe on public lands 95 percent of the time. In light of ongoing dwindling interest in hunting, perhaps hunters ought to be allowed on just 5 percent of public lands for just 5 percent of the time. Such changes in U.S. policies would truly hold U.S. public lands "in trust for the American people."

Public Lands

"Public lands are designated as held in trust for the American people," but < 5% of Americans, hunt and public lands are, nonetheless, still managed for hunters.

New Management Model:
Three Options

5%

For public lands to be held in trust for the American people, hunting should be restricted to:

- 5% of total public lands, or
- 5% of daylight hours, or
- 5% of daylight hours on 5% of public lands.

s3.7 Public Lands

Hunters are increasingly aware of their vulnerable status as an extreme minority, and they (and the gun lobby) have organized to preempt any curtailing of special treatment. For example, they have drummed up the Hunting Heritage Protection Act in the hope of placing "obstacles in the way" of anyone wishing to restore public lands to the general public ("Hunting Heritage"). Recently introduced in Congress, the Hunting Heritage Protection Act *requires* "federal wildlife agencies [to] manage public lands by promoting and enhancing recreational hunting and trapping opportunities" ("S. 1522 Hunting"):

> [F]ederal public lands would be managed so that there would be "no net loss of land area available" for hunting and trapping. These agencies would be required to submit annual reports detailing any federal lands that are closed to hunting and trapping as well as the reasons for the closure and the public lands that were opened to hunting and trapping to "compensate" for the closure. ("S. 1522 Hunting")

In short, the Hunting Heritage Protection Act, based on the disingenuous assertion that contemporary hunting is a U.S. heritage/tradition, seeks to seal into law the benefits that hunters have long enjoyed at the expense of other taxpayers. Unless common hunting hype is exposed for what it is, they are likely to succeed.

| Government Injustice |

- **Wildlife Services**—"manages" public lands and wildlife for hunters and ranchers (< 5%).

- **Pitman-Robertson Act**—taxes handguns purchased for personal protection; money spent on behalf of hunters.

- **Wildlife "refuges"**—wildlife on public lands is "managed" for hunters.

- **"Public" Lands**—ecosystems "managed," and thereby altered, on behalf of hunters.

(freeclipartstore.com)

s3.8 Government Injustice

Deceit of the Public

In a civilized and cultivated country, wild animals only continue to exist at all when preserved by sportsmen.

—THEODORE ROOSEVELT

Hunting hype is so common in the U.S. that it is difficult to discern the truth, but if we are to begin to make necessary changes on behalf of public lands, wildlife, and tax monies, these myths must be exposed for what they are, and debunked.

Hunting as "Tradition"/"Heritage"

Plain and simple, hunting is my heritage, and it's in the heritage of every American, and frankly no one is going to take that away from me.

— NORMAN SCHWARZKOPF QTD. in Scully 67

The presumption that "Man" evolved as a hunter has been challenged by recent anthropological theory, according to which humans have been foragers, not hunters, throughout most of our existence.

—LUKE, "A Critical Analysis" 7

Many people feel that some practices are protected simply because they have been around for a long time, but of course this is not the case. While in India, British general Sir Charles Napier forcibly ended the tradition of burning widows. When locals objected to this meddling with ancient customs, he replied: "This burning of widows is your custom; prepare the funeral pile. But my nation has also a custom. When men burn women alive we hang them, and confiscate all their property. . . . Let us all act according to national customs" (Napier 35).

Hunters sometimes defend their blood sports as "heritage" or "tradition," hoping to secure their favorite pastime in perpetuity. But as with bride burning, that which is considered "tradition" or "heritage" by powerful, self-interested individuals (who often exploit and harm others in the name of their coveted traditions), is subject to moral scrutiny, and may well be rejected and banned.

As with "refuge," defining terms is critical. "Tradition" and "heritage" refer to the same phenomenon: something handed down from the past, something that carries privilege of place, a certain historic entitlement. Definitions of tradition generally include three aspects: A "long-established or inherited way of thinking or acting," a "continuing pattern" of cultural beliefs or practices that are "handed down," and a "customary or characteristic method or manner," usually treasured by the larger community ("Tradition"). Traditions entail a practice, valued by the community, that maintains long-established methods and a concurrent inner quality—a mindset. Put simply, traditions are

- a *continuing* pattern of activity
- that employs specific *methods,*
- including specific *attitudes* and/or *ways of thinking,*
- all of which have been *handed down*
- because they are *socially meaningful.*

If we assume, just for the moment, that hunting is not inherently unjust and overall harmful (and therefore subject to termination) does contemporary hunting qualify as tradition or heritage? Have contemporary hunters maintained a continuing pattern of activities and methods across time—including a particular mindset—all of which are socially meaningful? Answering this question requires a sense of history, an understanding of methods used across time.

In Roosevelt's time men walked or rode a horse in order to track animals, carrying a simple rifle or shotgun, moving quietly to get within bullet range. Hunters also carried bodily remains on horseback, in a horse-drawn wagon, or (most often) on foot—dragging larger corpses. For example, my grandfather hunted on foot with a .22 bolt-action rifle (mammals) or a 16-gauge shotgun (birds). Though the practice was already illegal, he hunted with a dog who flushed target species out of thickets. If successful, he gutted the corpse in the woods and dragged or carried remains home.

The first 200 years of U.S. history saw the introduction of a variety of bullets and differing ways of loading ammunition—but these changes were slight compared with 20th-century innovations. Nineteenth-century rifles had "reasonable accuracy up to 600 yards" (550 meters—about a third of a mile or half a kilometer); some were multishot rifles ("Minié Rifle," "Rifle"). In comparison, 20th-century rifles kill from four times this distance, 1.5 miles (2.5 km), and come complete with scopes that magnify targets 9 times, and can fire a series of bullets in rapid succession. In open terrain, modern weapons easily drop a targeted individual from 1,000 yards (or more) ("Rifle").

- Rifles accurate for 1.5 miles

- Scopes magnify targets 9x

- Digital rangefinders (estimate distance between gun and target)

- Night-vision scopes (resolution beyond current military standards)

- Rapid-fire guns

- High-tech, high-powered bows

- ATVs

- GPS

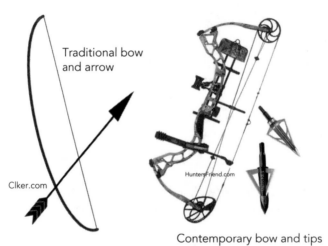

19th vs 21st century hunting rifles

- Attractants: decoys, hormones, sound imitation, etc.

- Infrared thermal imaging

§3.9 21st-Century Hunting Technology

Traditional bow and arrow

Clker.com

HuntersFriend.com

Contemporary bow and tips

§3.10 Traditional versus Contemporary Bows and Arrows

Clothing, footwear, and means of transport have also changed. For example, contemporary hunters rarely walk to their hunting grounds—in fact, they rarely hunt around their homes as my grandfather did. Contemporary hunters drive to hunting sites, and frequently use off-road vehicles to scout, hunt, and transport carcasses (Lewis and Paige). GPS systems, which require absolutely no navigational skills, have replaced compasses. Twenty-first-century hunting paraphernalia includes "chemical weapons" designed to mask human smells and/or deliver luring scents, including such products as

Wildlife Research Center Special Golden Estrus—bottled urine taken from "does brought into heat early through the use of hormones and lighting conditions"—sure to bring in a horny buck (Heffernan 25). Modern hunters can also purchase "autonomous, infrared-triggered trail cameras" for hunter surveillance, digital rangefinders to estimate the distance between the steel tip of their gun and the soft flesh of their target, and night-vision riflescopes promising "resolution beyond current military standards (Heffernan 25).

Hunters have not maintained a continuing pattern of activities or specific methods handed down from the past. Yet these are the least of the problems that face those who wish to consider hunting a "tradition"—a rightful heritage. The larger problem lies in *attitude, mindset,* and/or the *way of thinking* that is supposed to accompany a tradition. While Roosevelt felt that hunting channeled a man's virile impulses and helped turn men into gentleman, my grandfather—no doubt more typically—hunted in order to feed his family. Like many rural Americans in the early 20th century, my father remembers eating a lot of potatoes on the little eight-cow dairy farm where he grew up. Grandpa brought home deer, rabbits, ducks, pheasants, and (once) a bear. Grandma cooked up the flesh and if it was summertime, canned the meat to carry her family through lean months. Grandpa hunted in the winter out of need, even though it was illegal to do so. To evade detection, he simply hung carcasses in nearby woods, out of sight of the patrolling game warden. My father recalls that the fine for illegal hunting, if you were caught, was "steep"—but they had no cash, so any fine would have been more than they could have paid.

In contrast with hunters 100 years ago, people hunt today largely *because they enjoy the experience.* "In the United States, the conception of hunting as a pleasurable, recreational activity emerged in the middle of the nineteenth century in response to increased urbanization and leisure time" (Kheel, "Killing Game" 33–34). This altered purpose is reflected in contemporary ways of thinking about the hunt. Whereas my grandfather considered hunting a duty, a responsibility, a critical part of providing for his family, hunters today tend to view hunting as quality time with family and friends, or they hunt because they crave time away from the house and the daily workaday world—time in the great outdoors. Importantly, my grandfather had no other means of securing food for his family, while contemporary reasons for hunting can be satisfied in a variety of ways, such as hiking, camping, or canoeing.

There appears to be another motivation for hunting, a drive that is more difficult to satisfy through camping or canoeing. Less often articulated—but clearly an important reason for hunting—is the feeling of rugged individualism that urban folks crave when they escape their accustomed maze of roads, hand sanitizers, and flush toilets to seek and kill a wild animal, stick their hands into his or her corpse, scoop out viscera, toss the gutted body onto their vehicle, then cruise home to the suburban streets of their community, victorious. Indeed, we have lost touch with where our food comes from, with that tired and dirty feeling of having spent the day on basic tasks of survival. As a culture, we have lost touch with the earth and the natural environment in general, and hunting is one way that some people attempt to reconnect. But reconnecting with nature is certainly not a traditional reason for hunting—this is not the hunter's heritage. Further, though "many sport hunters claim to hunt to achieve intimacy with nature," this method does not seem likely to succeed. In the strong words of Marti Kheel, "[J]ust as the rapist does not achieve genuine intimacy through rape, . . . hunters do not achieve genuine intimacy with the animal that they kill ("Killing Game" 39). Surely there are more likely ways to connect with nature than needlessly killing wildlife. Moreover, we do not generally gain a sense of oneness with those whom we choose to dominate and destroy.

For those genuinely interested in reconnecting with the natural world, killing wildlife with high-tech weapons does not seem the most obvious choice. In truth, many people hunt because they get a thrill out of killing. Taking a life is often described as an exciting moment, a moment that seems to be difficult to describe with words—an experience that some men describe as orgasmic. Killing without need—the thrill of killing for sport— "includes a specifically sexual component" (Luke, *Brutal* 107). "Within patriarchal social structures the disposition to take sexual pleasure in the domination and destruction of other living beings is a normal part of men's fulfillment" (Luke, *Brutal* 87). While no doubt killing is very exciting for sport hunters, taking a good look at "the normalcy of men's erotic enjoyment of hunting suggests the threatening possibility that there is something seriously wrong with normal manhood in this culture" (Luke, *Brutal* 87). Again, aside from this very serious concern regarding manhood in our larger culture, sexual fulfillment is certainly not a traditional reason for hunting—this is certainly not why my grandfather trudged across the snow in subzero temperatures in search of a rabbit or grouse. Nor is such an "orgasmic" experience meaningful to the larger society.

Until about fifty years ago, U.S. citizens hunted out of necessity, with low-tech, short-range guns, often seeking and retrieving wildlife on foot. Hunters have *not* maintained a *continuing* pattern of activities, using specific *methods* that coincide with certain *attitudes* and/or *ways of thinking, handed down* because they are *meaningful to the larger society.* Contemporary hunting is not a "tradition" and does not qualify as "heritage." On reflection, gardening qualifies much more easily. It appears that we evolved largely (if not almost completely) as gatherers; perhaps gathering is the rightful heritage of humanity (Luke, "A Critical Analysis" 11). In any event, gardening provides an excellent way for families to share time, reconnect with nature, and produce low-cost, nutritious foods.

Contemporary hunting, as practiced by the vast majority of U.S. hunters (including those who wielding high-powered, high-tech bows and arrows—a hunting choice that would have baffled my grandfather), does not qualify as tradition or heritage. It seems critical to scrutinize the popular practice of sport hunting specifically:

> "Sport" is typically conceived as a voluntary agreement among participants, complete with mutually understood rules and goals. But other-than-human "players" in this "game" do not consent to the rules, or even to be competitors. The hunter sets the rules, including the goal to kill; animals, if aware of a hunter's presence, can only respond by attempting to flee. The forcible inclusion of other-than-human animals into this "game" renders meaningless the analogy with sport and the idea of "fair chase." (Kheel, *Nature Ethics* 91–92)

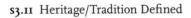

- a *continuing* pattern of activity,

- that employs specific *methods*,

- including specific *attitudes* and/or *ways of thinking*,

- all of which have been *handed down*,

- because they are *socially meaningful*.

(clker.com)

S3.11 Heritage/Tradition Defined

Even if modern hunting qualified as "tradition" or "heritage," it is likely that this sport would be cast into that dark pit of history, along with other ancient and barbaric practices such as bride burning.

Hunting as Natural

Some argue that hunting is "natural," presumably referring to predator/prey relations, to the fact that killing and dying are part of life. Indeed, we cannot live without killing, and it is therefore sometimes argued that hunting is no better or worse than any other means of acquiring sustenance. Is this true?

The claim that we live in a dog-eat-dog world ignores the earth's innumerable herbivores—antelope, rabbits, elephants, vegan humans, mice, elk, hippopotami, moose, caribou, buffalo (and on and on)—who need not kill to live, and the blessing of choice that is likely for those reading this book. Herbivores, including the earth's largest and strongest animals (elephants, bison, etc.), select leaves and nip the ends off of grasses. While killing is a way of life for some species, it is clearly not essential for human health. We are not designed to run down deer or squirrels and sink our teeth through hair and hides to reach and devour muscle tissues. We do not have short intestines, flesh-tearing fangs, or dangerous claws: We have long intestines and grinding molars, like other plant eaters. It should not surprise us that the most common human medical problems in the developed world are linked with a diet rich in animal products. Our digestive tracts and teeth are similar to those of other herbivores, not those of omnivores (like bears and coyotes) or true carnivores (such as bobcats and cougars).

Those who argue that we ought to tolerate hunting (and meat eating more generally) because killing is an inevitable part of life, also seem unaware of the notion of ethics. Such an argument requires that we also tolerate murder and rape because they have always existed (and likely will always exist). Morality takes for granted that we have a measure of choice in how we live, that we can and ought to discourage destructive and harmful behaviors in ourselves and others. Perhaps more to the point, outside of defending hunting (or omnivory more generally), when do human beings look to *other* species to decide how *we* ought to behave? Why—on this particular point and no other—do we look to other animals to determine what is morally acceptable?

Argumentative omnivores also sometimes suggest that animal advocates are obligated to prevent predators from killing prey if such advocates sincerely believe that killing to eat is morally problematic. This assertion is specious. A morally relevant distinction, namely biological necessity, separates those who must hunt to live (wild carnivores) and those who sometimes hunt for pleasure (roughly 5 percent of U.S. citizens, for example). To deprive a carnivore of prey is to deprive him or her of life itself, and the same is true, to a lesser degree, for wild omnivores. To deprive a sport hunter of prey does not leave him or her hungry—and if it does, such need-based hunters do not fall under the moral scrutiny of this chapter, as noted at the outset.

Hunting as "Mercy Killing"

Wildlife "managers," being both delegates and lackeys for the hunting industry, would have you ingest the preposterous pabulum that hunting helps animals; that hunters are their philanthropic fairy godparents (well-armed well-wishers, if you will) performing a gallant duty; that animals won't go on living unless we kindheartedly kill them. This is all the more outrageous in light of how many species have been wiped off the face of the earth, or perilously close to it, exclusively by hunting.

—ROBERTSON 13

The dominating myth, of course, is that wildlife are overpopulated and in need of culling, a silly notion considering that humans are really the only animals that are overpopulating.

—MOSS

Aldo Leopold touted hunting as an "expression of love for the natural world" ("A Leopold Biography"). Contemporary hunters and many employees of FWS continue to assert that shooting wildlife is beneficial—even essential—for wilderness, ecosystems, and wildlife populations. FWS even goes so far as to suggest that hunters protect human health and welfare:

> [H]unting is an important tool for wildlife management . . . a valuable tool to control populations of some species that might otherwise exceed the carrying capacity of their habitat and threaten the well-being of other wildlife species, and in some instances, that of human health and safety. ("Hunting")

Of course FWS does not bother to explain details with regard to hunter-target species posing a threat to human health and safety—or admit to human responsibility for species imbalances. FWS merely presents these unsubstantiated claims in order to tout hunting as a *necessary service* to humanity, ecosystems, and wildlife.

Based on such propaganda, many contemporary U.S. citizens continue to believe that in the absence of hunters, deer and elk would overrun the planet, wreak ecological havoc, then slowly starve. Ironically, FWS (and hunters) have simultaneously convinced many citizens that coyotes and wolves would similarly swarm the planet without hunters (and trappers) to keep them in check. But this cannot be true for both predator *and* prey. Didn't we all learn in biology that predator/prey relations have a measure of cyclical flux that is predictable and stable. . . if left alone?

Deer are the most common species referenced by hunter advocates with regard to overpopulation and starvation—no one wants Bambi to die a slow and painful death in the dead of winter. In truth, hunting is the problem with regard to ungulate starvation, not the solution. For example, wildlife agencies "manipulate the ratios of bucks to does" through hunting licenses "in an attempt to yield the maximum number" of hunter-target species (Yarrow). Sex ratios are thereby deliberately skewed to favor life-giving does (Yarrow). If deer are truly a population hazard, and overpopulation is, in fact, why hunters shoot deer, then why are hunting licenses designed to *enhance* deer populations? Targeting females would reduce deer populations much more efficiently—fewer ungulates is clearly not the intent. Moreover, if deer are dangerously overpopulated, why is hunting limited to certain seasons, and why are kill numbers limited? It is disingenuous to argue that deer must be hunted *because they are overpopulated*, while fostering policies designed to *inflate* deer numbers. More importantly, if deer (and other hunter-target species) are at risk of overpopulation and starvation, shouldn't FWS stop eliminating (and start protecting) their natural predators? If hunters are sincerely concerned about deer populations, why are they "the first to resist" the reintroduction of natural predators (Scully 66)? Finally, if hunters are sincerely concerned about wildlife overpopulation and starvation—for prey or predators—why do they shun immuno-contraceptives that have proven 80 percent effective?

Those who claim that hunters prevent slow death by starvation also overlook the fact that hunters do not target thin, vulnerable, or sickly deer. Hunters tend to seek big bodies and big antlers—the bigger the better. Rather than eliminate sickly deer, hunters strip "big game" populations of their strongest members, and their strongest genes. "This sort of discriminatory culling-of-the-fittest runs counter to natural selection and is effectively triggering a reversal of evolution by giving the unfit and defective a better shot at passing on their genes" (Robertson 123). If hunters were concerned about slow death by starvation, they would use the hunting tag they purchase to "take" a small, thin moose, elk, or deer.

Perhaps most remarkably, those who argue that hunting provides essential population control never seem to notice the hypocrisy—the absurdity—of human beings gunning down other species because we think *they* are overpopulated. Do we really want to advocate shooting-to-kill as a reasonable and appropriate solution to overpopulation? This seems a bit foolish given our ever-growing, environmentally damaging numbers. The assertion that killing is an honorable solution to starvation seems particularly problematic in light of mass human starvation and chronic hunger in various places around the world. Moreover, there can be no doubt that humans, more than any other species, have exceeded the carrying capacity of the planet, and that our excessive fertility threatens "the well-being of other wildlife species, and in some instances, that of human health and safety" ("Hunting"). But killing those who are hungry—whether ungulates or great apes (human beings)—is not a solution that most of us are willing to condone.

In any event, hunter-target species tell a different story. Of the whopping 200 million animals killed by U.S. hunters each year, just 4 million are deer (Gudorf and Huchingson 251). Most hunter-target species must be and are *protected* by FWS (including deer in many areas, as noted above). To argue that hunters kill wildlife out of benevolence for those they kill is clearly disingenuous. Such boldfaced hypocrisy can easily be avoided by accepting that almost all contemporary North American hunters hunt because they enjoy hunting, and for no other reason. Perhaps that is the problem—hunters don't want to admit that they *enjoy* shooting Bambi's mother.

Revisiting Deer Overpopulation Myths

(WPClipart)

- Deer are *protected* by FWS policies:
  - Sex ratios are purposefully skewed to increase numbers.
  - Hunting seasons limit take to protect populations.
  - Natural predators are systematically reduced to protect deer.
- "Culling-of-the-fittest:"
  - Fails to remove sickly or weak.
  - Harms genetic evolution, removing biggest and best.
- Shooting is not generally consider a reasonable way to remedy overpopulation or hunger–
  - What about immunocontraceptives are?

Hunting is not generally a means of population control–200 million animals are shot by hunters annually, but only 4 million of these are deer. Many hunter-target species must be carefully monitored to maintain numbers.

s3.12 Revisiting Deer Overpopulation Myths

Hunting as Procuring "Compassionate Meat"

In recreational hunting and angling, animals are classified as moving targets for those who enjoy killing or capturing others.

—LUKE, *Brutal 2*

[I]f hunters were truly attending to nature, instead of to their own amorphous feelings . . . they would feel the terror and fright of the animal they seek to kill.

—KHEEL, "Killing Game" 40

With heightened awareness of the horrors of factory farming, some argue that hunting is preferable to buying beef or pork or turkey—that hunting elk or rabbits or quail is more compassionate than buying factory-farmed animal products: "Many hunters point out that death in the slaughterhouse is by no means more humane" than a bullet in the wild (Luke, "A Critical Analysis" 6). Similarly, when consumers learn of the prolonged and intense suffering of cows and calves for the sake of dairy products, and of hens for the sake of eggs, hunting appears less objectionable.

Unfortunately, hunting also causes much misery. Hunters cannot and do not shoot perfectly each time they pull the trigger. Most hunters are only in the field five to seven days per year—hardly enough experience to secure the likelihood of a clean shot. Novices, bent on achieving their legal limit, often "shoot from the hip" and injure rather than kill their target (Gudorf and Huchingson 252). Hunters frequently leave wounded animals (especially birds) to die slowly in the elements, or to be harassed and killed by predators. Hunter wounding without retrieval is estimated to be as high as 30 percent (Gudorf and Huchingson 252; "Reducing").

Of the 200 million animals that are killed by hunters every year, 50 million are doves, 25 million are quails, and 20 million are pheasants (Gudorf and Huchingson 251). Fowl are gunned down with shotguns that scatter pellets, the same method that my grandfather used. Pellets randomly hit wings, tails, and various portions of the stomach or legs, but are not "aimed" at vital organs. Such scatter-shot may or may not immediately knock birds out of the sky; they are very unlikely to bring quick death. Based on likelihood of random pellets hitting the head or heart directly, most birds tumble from the sky still alive. Hunters frequently use "bird dogs" in the hope of retrieving those wounded, but birds with leg injuries or flesh wounds are unlikely to fall within a dog's range, and many wounded animals are never retrieved, and die from infection, blood loss and shock, predator attacks, starvation, and so on.

Even those who are retrieved suffer before they die. For birds, this suffering includes falling from the sky, realizing they cannot escape because their wings or legs will not function, being approached and taken into the mouths of dogs, and then delivered into the hands of a hunters who wrings their neck (Hatfield 6).

- 200 million animals shot annually
- Almost half of them are fowl:
  - 50 million doves
  - 25 million quails
  - 20 million pheasants
- Wounding rate for fowl—30%

(Clker.com)

Statistics show that quick, clean, single-shot kills are uncommon among hunters.

s3.13 Do Hunters Deliver a Quick Death?

I frequently hear from bow hunters that they have moved away from guns in search of a challenge. Indeed, with long-range rifles and abundant ungulates, hunting is remarkably easy. Indeed, bow hunting is more difficult, and despite the use of high-tech bows and arrows, bow hunting has a wounding rate of greater than 50 percent. *Bow hunters expend an average of 14 arrows per kill*, and "[w]ounding and crippling losses are inevitable" (Hatfield 1). Each of 24 studies on the subject "concluded that for every deer legally killed by bowhunters, at least one or more is struck by a broadhead arrow, wounded, and not recovered" (Hatfield 1).

Wounding rates only take account of physical injuries, ignoring psychological wounds and psychological suffering, despite the fact that mental suffering is often recognized as more intense and unbearable than physical suffering. Imagine the psychological terror of a deer or elk—both herd animals—when members of their community suddenly fall over and begin kicking and struggling for life, without any sensory evidence that there is a predator in the area. Those who live in hunting areas (myself included) can readily see the change in stress levels of hunter-target species during hunting season. In areas where hunting is never allowed, deer

- Wounding average: 54%

- Average arrows per kill: 14

- *Each of 24 studies concluded:* "For every deer legally killed by bow hunters, one or more is . . . wounded and not recovered" (Hatfield 1)

(Clker.com)

s3.14 Bow Hunting: Wounding and Crippling Are Inevitable

do not fear humanity; prior to hunting season, deer are visibly relaxed, watchful but not edgy as they graze or nap on open landscapes. With the advent of hunting season, they are rarely visible, and bolt at the slightest sound. Hunting instills dread and terror in target animals—mental pain and suffering (along with physical pain and premature death): Hunters leave "victims emotionally scarred" (Robertson 134).

It would seem fairly obvious that sport hunting is cruel, but to be sure, we must review the definition of "cruel":

- "disposed to inflict pain or suffering,"
- an act "causing or conducive to injury, grief, or pain." ("Cruel")

Indeed, hunting is cruel. Small wonder that "fully twenty percent of Americans who hunt in their youth give it up because they become convinced that it is wrong"; including the many hunters who continue to travel with buddies to their favorite hunting grounds but never again bring home a dead animal (Luke, "A Critical Analysis" 13).

To determine whether shopping for flesh or hunting is morally preferable, it is also necessary to take into consideration the physical and psychological pain and premature deaths of the many animals gassed, trapped, snared, and gunned down on behalf of hunters for the sake of predator control. Snaring and trapping are known to cause extensive, prolonged suffering not just for those caught but for their larger communities, whose members often watch their family and friends as they languish in a trap until a human being comes to kill them. Humans are also traumatized when they stumble on a prairie dog or coyote, terrified and suffering, with a leg caught in a trap.

There is no nutritional need to consume flesh—venison or beef or chicken. People who are genuinely committed to minimizing suffering must ask a broader question: Do I need to eat animal products? Those who sincerely wish to reduce suffering (and protect ecosystems)—will hunt out potatoes and pickles rather than deer and ducks. For those sincere in their quest for a compassionate diet, the answer is vegan, not venison.

Hunting for Economy

Hunting is no longer motivated by hunger. Twenty-first century sport hunters are never without a full belly, even after investing tens of thousands of dollars on brand-new 4x4 pickups, motorboats, RVs and of course the latest high-tech weaponry.

—ROBERTSON 85

Many hunters justify their sport as a means of low-cost sustenance, but when a prospective hunter inquired online what it might cost to get into the game, replies noted the expense of a rifle/bow, ammunition, boots and clothes, license and permit—$500 to $1500 just to get started ("How"). This response did not include the need for a chest freezer and a truck—or some means of transporting bloody carcasses from rough back roads to the kitchen—and the possible cost of professional butchering. Not much of a markdown meal—and these are just start-up costs (Figure 3.1).

What is hunting likely to cost in a given year? Ongoing expenses include ammunition, annual hunting licenses, the cost of transport, and gear upgrades and replacements. At $.50–1.50 for a single shot, ammunition is expensive (Gudorf and Huchingson 252; Robertson 85). (At this cost, how many hunters invest in honing hunting skills at a target range before they head for the woods?) Seasonal hunting licenses in the U.S. range from $20–50 (depending on the state), with an additional license (in the range of $10–30, again depending on the state) required to shoot certain hunter-target species, such as a deer; even higher fees must be paid for the privilege of killing a less common species (elk, antelope, moose, for example). Hunting out-of-state generally costs at least four times as much . . . and hunting out-of-state is increasingly necessary as urban sprawl claims local wildlands ("How Much"). Indeed, many hunters drive long distances to reach areas open to hunters, and travel has become an increasingly significant portion

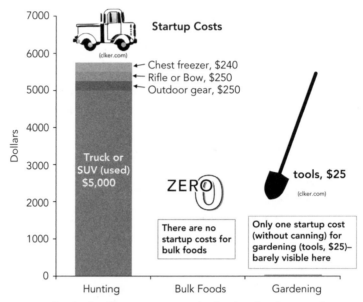

FIGURE 3.1 What is the Cheapest Way to Feed a Family: Startup Costs

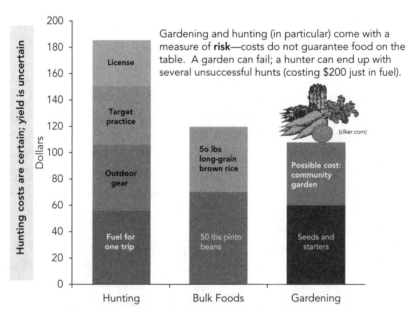

FIGURE 3.2 What is the Cheapest Way to Feed a Family: Annual Costs

of the cost of hunting (along with owning, insuring, and maintaining an appropriate vehicle for hauling corpses across rough terrain, or renting such a vehicle as needed). This does not equate to a bargain buck (or even a discount duck) (Figure 3.2).

In comparison, consider gardening—a healthy enterprise that puts people in touch with their food and carries innumerable opportunities for bonding, and is considerably more economical. Seeds and starters ($5–$15, depending on amount and variety) are the primary annual cost. Watering a garden entails no additional expense for those who already water a yard. For those who do not have a yard, community gardens are available in many locations (often free, sometimes $15–50 for a season, depending on the community) ("Frequently Asked Questions"). It is possible to can or freeze produce, but gardeners can also choose to plant crops that keep well, such as winter squash and potatoes. Gardening is much cheaper than hunting, especially if one adds environmental costs and the cost to one's long-term health.

Another option, also much more economical than hunting, is to buy bulk staples (Figure 3.3). For example, it is pretty easy to find 20 pounds of rice for

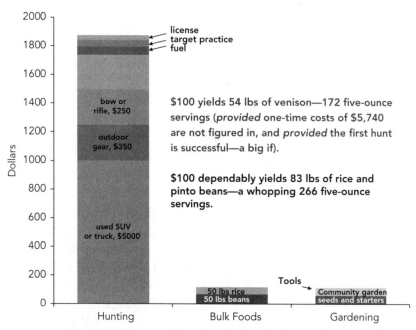

FIGURE 3.3 What is the Cheapest Way to Feed a Family: Startup Plus Annual Costs

| Comparing Costs |

Minimal One-Time Costs of Hunting:
- Carcass-carrying vehicle, used, $5,000
- Bow or rifle, about $250
- Outdoor gear, $250
- 24-cubic-foot chest freezer, $240

Minimal Annual Costs (w/out butchering)
- Fuel for transport, 80 miles roundtrip @ 20 mpg @ $3.50 per gal = $56
- Outdoor gear, $50
- Target practice, $.70 per shot, 50 shots per season: $35 plus $10 fuel = $45
- Basic seasonal U.S. hunting license $20-50

Average lbs eatable meat per deer = 100; 3 servings per pound (5-ounce servings) = 300 5-ounce servings.

$100 yields 54 lbs of venison—enough for 172 five-ounce servings, provided one-time costs ($5,740) are not figured into the equation.

Minimal One-Time Costs of Gardening:
- 40 quart jars, rings, lids, $55
- Canner, $100
- Tools, $25

Minimal Annual Costs:
- Seeds and starters, $5-15
- Community garden plot, $15-50
- Lids $5%

Minimal Annual Costs of Buying Bulk
- Fifty lbs. of rice, $50 1 lbs = 2.5 c dried = 5 c cooked; with a serving size of .5 c (3.5 oz.), 50 lbs = 250 c cooked = 500 meals

- 50 lbs of pinto beans, $70 (1 lb = 2 c dried = 6 c cooked, 2 c per lb, 2/3 c uncooked yields 1.5 c cooked, l/2 c (4 oz) serving size—50 lbs = 300 c cooked = 600 meals

$100 yields 83 lbs of rice and pinto beans—enough for 266 five-ounce servings.

(freeclipartstore.com and wpclipart.com)

FIGURE 3.4 Comparing Costs

$10, and 24 pounds of navy or kidney beans for even less. Much of the world's population has depended on these, or similar grain- and legume-based staples for centuries—beans and rice, rice and lentils, potatoes and beans. With bulk grains, legumes, and plants, it is easy to prepare delicious meals for one person on a budget of $20–30 per week—including spices—now *that* saves bucks (Figure 3.4)! (See Ellen Jaffe Jones' book *Eat Vegan on $4 a Day.*)

When sums are calculated, hunting is clearly a comparatively expensive way to put food on the table: "In reality, hunting is typically not a source of provision but actually drains family resources. Deer hunters, for example, spend on average over forty dollars per pound of venison acquired, once all the costs of equipment, licenses, transportation, unsuccessful hunts, and so forth, are calculated" (Luke, *Brutal* 88). In a bow-hunting magazine, a hunter admits: "Nobody hunts just to put meat on the table because it's too expensive, time consuming, and extremely inconsistent"; hunters kill wildlife "because it's fun!" (Luke, *Brutal* 88). Honesty, while sometimes disturbing, offers much-needed clarity. North Americans hunt for pleasure, not for sustenance. Perhaps the best proof of this is that hunters kill some twenty-five million mourning doves, sixteen million squirrels, two million woodchucks, half a million prairie dogs, six hundred thousand crows, and sixty one thousand skunks annually ("Learn" 2). Anyone insisting that hunters consume all that they kill must eat crow, or perhaps something even less palatable. Skunk stew, anyone?

Hunting and "Fair Chase"

Deer hunting would be fine sport, if only the deer had guns.

—WILLIAM S. GILBERT, from "Quotations about Animal Rights"

Hunting is not a sport. In a sport, both sides should know they're in the game.

—PAUL RODRIGUEZ, from "Quotations about Animal Rights"

Perhaps the most disingenuous aspect sport hunting is the deceit of "fair chase," a concept that forms "the cornerstone" of U.S. "game" laws and lies at the core of the Western hunting ethic ("History"; "About the B & C Club"). "Fair chase" is envisioned as the "ethical, sportsmanlike, and lawful pursuit and taking of free-ranging wild game animals in a manner that does not give the hunter an improper or unfair advantage" ("History").

By way of analogy, a "fair fight" involves opponents who are *equally matched* and who *choose* to engage in battle under a particular set of rules. For example, if one combatant is much smaller or younger, that combatant is disadvantaged, and the fight is considered unfair. If one combatant has a wooden bat and the other has a .45-caliber automatic pistol, the fight is considered unfair. Or, if one person ambushes an unsuspecting passerby, the ensuing fight is not considered fair.

"Fair chase," if we are to use English without grossly distorting the meaning of words, ought to be similar. By what stretch of the imagination does a deer *willingly* become a hunter's target? Is an armed man a fair match for an unarmed duck? How can the stakes be equal when a hunter choses to come to the field for recreation—for fun—while the "game" animal has no interest in being hunted, yet may or may not be wounded, maimed, or killed in the course of the hunter's sporting adventure?

Because today's North American hunters are generally not facing starvation if they find no success in the hunt, they have the luxury of drumming up "rules of fair chase to make the hunting experience a more exciting challenge"—and to assuage moral concerns/guilt that accompany such an inherently unfair sport (Luke, "A Critical Analysis" 9). "Hunting and fishing involve killing animals with devices (such as guns) for which target animals have not evolved natural defenses. No animal on earth has adequate defenses against a human armed with a gun [or] a bow and arrow" (Bekoff 121). By what stretch of the imagination would it be possible for a hunter toting modern weaponry *not* to have an "unfair advantage" over geese, elk, bears, gophers, and quail? In the absence of an "improper or unfair advantage"—by definition—roughly 50 percent of hunting excursions would result in hunter deaths at the "hands" of their equally matched, read-to-kill "opponents." If hunting entailed "fair chase," I suspect there would be precious few hunters.

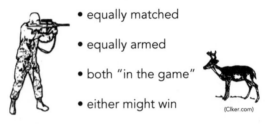

- equally matched
- equally armed
- both "in the game"
- either might win

(Clker.com)

If hunting entailed "fair chase," there would be precious few hunters.

§3.15 What Is "Fair Chase"?

In truth, hunting does not entail *any* chase. It is extremely difficult to shoot fleeing wildlife, and it is pretty much impossible for a running, gasping hunter to successfully aim and fire a gun. Hunters are therefore most likely to fire at wildlife who have not taken flight, who have likely not even noticed the hunter's presence (which is easy to accomplish with today's long-range, precision weaponry). Instead of engaging in "fair chase," hunters conceal their presence in the hope of killing hapless wild animals as they graze, nap, or scan the horizon for predators. This better fits the definitions of:

- "ambush"—attack from a concealed location ("Ambush");
- "massacre"—"killing a number of usually helpless or unresisting" victims, including the "wholesale slaughter of animals" in a way that is "cruel or wanton" (especially apt for bird hunting) ("Massacre"); and
- "assassination"—"to injure or destroy unexpectedly and treacherously" ("Assassinate").

Regardless of the artifice of "fair chase," many hunters seek an even greater "improper or unfair advantage" through "advanced electronics, weaponry, chemicals, and camouflage, all designed to eliminate every last shred of chance . . . despite the fact that finding a deer to kill has literally never been easier" (Heffernan 25). Hunters, like many other contemporary people in industrialized nations, seem obsessed with gadgetry.

If we look honestly at hunting, "fair chase" is a misleading incongruity concocted in the hope of retaining the pleasure of the hunt as a legitimate form of recreation, despite the fact that such killing is completely unnecessary and embarrassingly easy. "Fair chase" is a lie designed to cover up what is, in truth, an ambush and a massacre.

- **Ambush**—attack from a concealed location
- **Massacre**—"killing a number of usually helpless or unresisting" victims
- **Assassinate**—"to injure or destroy unexpectedly and treacherously"

(Clker.com)

s3.16 If the Shoe Fits . . .

The "Gentleman Hunter"

[T]here is a moral problem entailed in the idea of pursuing the death of another living being for the opportunity it affords one to engage in an enjoyable experience.
—KHEEL, "Killing Game" 32

These enthusiasts often like to hang signs that say "Gone Fishin'" or "Gone Huntin'." But what these slogans really mean is "Gone Killing."
—BEKOFF 120

Roosevelt asserted that hunting helped young men transcend their dangerous passions, that killing wildlife *civilized* men, and for this reason he advocated hunting as a tool for moderating deadly urges and fostering mature young men—not as a means of feeding families. But who among us has met the "cultivated killer," the noble man who takes pleasure in killing in and of itself? Eating flesh is unnecessary for human health (and is more apt to damage health). How plausible is it that engaging in the act of needless killing creates "civilized" human beings?

Perhaps Roosevelt's error stemmed from his uncritical acceptance of "fair chase." Maybe he actually came to believe that sneaking up on an unarmed individuals, and shooting them before they knew you were there, could be construed as "fair chase." But it is obvious that contemporary hunters using modern devices hold tremendous advantage over hunted animals—and that they do so not out of need, but as a matter of entertainment, for pleasure, as something to do in their spare time. By definition, gentlemen do not kill for pleasure, entertainment, or sport—not even for personal gain. Recall the definition of cruel: "disposed to inflict pain or suffering," whereby a cruel act causes or is conducive "to injury, grief, or pain" ("Cruel"). Hunting is a perfect example of cruel behavior. Hunting is therefore unlikely among gentlemen, but perhaps likely among those who are indifferent to life and the suffering of other beings. "Gentleman hunter" is an oxymoron.

It is widely documented that those who pose a danger to defenseless animals had best not be trusted with the lives of defenseless human beings. Many a serial killer began his cruel exploits by torturing and killing nonhuman animals, including Ted Bundy, David Berkowitz, Jeffrey Dahmer, and the Boston Strangler ("Link between Animal Abuse"). One study found that among "36 convicted multiple murderers . . . 46% admitted committing acts of animal torture as adolescents. And of seven school shootings that took place across the U.S. between 1997 and 2001, all involved boys who had previously committed acts of animal cruelty" ("Animal Cruelty"). "Researchers as well as FBI and other law enforcement agencies nationwide have linked animal cruelty to domestic violence, child abuse, serial killings and to the recent rash of killings by school age children" ("Link between Animal Abuse"). Apparently, sometimes those who enjoy killing other animals have trouble stopping at the species barrier. While it would be disingenuous to suggest that all hunters are serial killers in the making, it would be equally disingenuous to ignore the documented link between those who are cruel to other animals and those who are cruel to other humans. It is foolhardy to overlook the well-documented link between animal cruelty and those at risk of using guns against human beings, especially in the U.S., a nation known for gun violence, a nation where guns are readily available, and where government agencies (FWS) encourage and foster hunting even among children.

In the U.S., as elsewhere, hunting is just another expression of power and control, another means of dominating and exploiting those who are comparatively weak and unable to defend themselves against those who hold power. A violent expression of power and dominance tends to be associated with males. In many cultures the "essence of manhood is the ability and willingness to destroy others" (Luke, *Brutal* 108). As a society, we must work to change these underlying, outdated, damaging vestiges of primitive manhood if we are to end the violence and oppression that ultimately harms all of us (Luke, *Brutal* 108). Dressing gratuitous violence in the trappings of "gentlemen hunter" lends legitimacy to cruelty, camouflaging hunters in a cloak of outdated myths.

Hunters are now an extreme minority, and they must work hard to maintain outdated, costly, environmentally damaging hunting privileges. Any attempt to portray hunters as benevolent is dishonest. Depicting hunters or hunting as a man's domain, as exemplifying "true manhood" and an expression of manliness—is damaging to men and dangerous for all of us. It is time to expose sport hunting as unmitigated violence, as cruel and completely unnecessary. But hunters are busy bolstering hunting myths to instead of their cruel sport in the hope of maintaining government-funded privileges. Hunters are keen to present hunting not only as a legitimate sport, but as a worthy enterprise and as a citizen's right. Today, in light of the hunter's "politically vulnerable minority status," many hunters are desperate "to secure hunting privileges and access to public lands, and they feel that to do so requires curbing behaviors that most blatantly display disrespect for animals" (Luke, "A Critical Analysis" 12). Deception remains critical if the majority are to continue to accept and support hunting.

Today, as in Roosevelt's time, hunters feel compelled to pass themselves off as civilized—as gentlemen. Today's hunters continue to foster hunting myths, hyping hunting as the perfect leisure activity, a manly sport bound by strict rules of sportsmanship, providing for families, a legitimate way to procure food and to spend time with friends and family while fostering conservation and preventing gentle deer from the horrors of starvation. "Gentleman hunter" is the epitome of hunting hype.

The "Eco-Hunter"

I don't apologize for being a hunter because hunters have done more to protect and increase animals and habitat than any other organization or group.

—NORMAN SCHWARZKOPF QTD. in Scully 67

Most people conceive of the environmental movement as designed to curb or eliminate our society's destructive relation to the natural world. It may, therefore, seem puzzling

to some that a growing number of environmental writers have endorsed an act of violence—namely, hunting.

—KHEEL, "License to Kill" 86

Aldo Leopold hyped hunting as an "expression of love for the natural world" and as a primary reason to preserve wilderness, thereby creating a channel through which "environmentalists" have aligned with hunters ("A Leopold Biography"). Many hunters and many U.S. federal and state wildlife agency employees continue to assert that conservation exists—and wildlife remains—thanks to hunters. While there was some truth to this when wildlife management first began, it is no less true that *wildlife management was only necessary because of hunters*. Hunters in North America eliminated helpless great auks, vast flocks of passenger pigeons, magnificent eastern elk, and gentle Steller's sea cows for personal pleasure, profit, and much less often (or incidentally), for food.

U.S. wildlife conservation—preserving and protecting "big game" for sport hunters—was born when hunters realized that their "game" would be up if they did not protect wildlife from their own guns. Consequently, hunters sought an overarching, well-funded system of management. A handful of influential early hunters held government positions through which they were able to tap into the powers of the U.S. government. To this day, federal and state government agencies manipulate wildlife on behalf of hunters (though not only at the expense of hunters). This outmoded, environmentally damaging form of conservation—government-funded manipulation of wildlife in order to conserve target species for hunters—continues to the present, and lies at the core of the vacuous claim that hunters are environmentalists.

Unfortunately, environmental organizations have done nothing to debunk this claim, and have instead courted hunters. Naturally, hunters are drawn to organizations that protect and preserve their interests, and many environmental organizations have done just that, "seduced by a desire to engage as many paying members" as possible (Robertson 131). Through their affiliation with hunters, contemporary environmental organizations continue to propagate the myth of the "environmental sport hunter" (Kheel, *Nature Ethics* 96). For example, the World Wildlife Fund, the National Wildlife Federation, the Sierra Club, the National Audubon Society, the Izaak Walton League, and the Wilderness Society actively appeal to and work on behalf of hunters ("Why Sport Hunting"). The Sierra Club recently held an essay contest on the subject "Why I Hunt"—offering the winner a trophy hunt in Alaska (Robertson 134).

Finances and affiliations aside, definitions are fundamental for deciding whether or not sport hunters qualify as environmentalists. An environmentalist is

- "a person who is concerned with the maintenance of ecological balance and the conservation of the environment";
- "a person concerned with issues that affect the environment, such as pollution";
- "an expert on environmental problems";
- "a person who advocates or works" to protect "air, water, animals, plants," and other aspects of the natural environment ("Environmentalist").

Even when hunters hold "expert" knowledge of environmental problems, or engage in advocacy on behalf of "air, water, animals, plants," or against "pollution," the first definition disqualifies *all* hunters (at least in the U.S.) from claiming to be environmentalists. Hunter dollars are used to manipulate wildlife in order to bolster hunter-target species, thereby altering ecosystems, and hunters therefore cannot reasonably claim to be "concerned with the maintenance of ecological balance," and therefore do not qualify as environmentalists. By definition, environmentalists seek to allow predator/prey relationships to create and recreate their own natural balance, as they have for millions of years. Conflicting with the environmentalist's agenda, the policies of Fish and Wildlife Services are designed to create bloated numbers of hunter-target species and a dearth of carnivores. Moreover, hunters continue to enlist powerful government wildlife agencies to manipulate predator and prey numbers—to manipulate ecosystems—on behalf of hunters. While nonhunters unwittingly pay the bulk of the costs for this work, it is widely known that hunter and fisher licenses support FWS wildlife manipulation.

One who is concerned with the maintenanceof ecological balance.

Ref: ("Enviromentalist")

(Kaboodle.com)

s3.17 Environmentalist Defined

Those interested in "the maintenance of ecological balance" will have nothing to do with hunting (or fishing). Indeed, environmental organizations that pass the plate to hunters (and fishers) sell out on natural ecosystems.

According to Aldo Leopold, a "thing is right when it tends to preserve the integrity, stability, and beauty of the biotic community. It is wrong when it tends otherwise" (Leopold, *A Sand County Almanac* 224–25). On behalf of hunters, wildlife agencies damage the integrity and stability of ecosystems while manipulating biotic communities. According to Leopold's principle, contemporary hunting in the U.S. is wrong: Hunting damages "the integrity, stability, and beauty of the biotic community" (Leopold, *A Sand County Almanac* 224–25). Contemporary environmentalists in the U.S. who align with or support hunters and hunting do so at the expense of their mission—and therefore at the expense of their own integrity.

Conclusion

Hunter-funds fall short of the costs of FWS wildlife manipulations, but instead of shifting from special interests to public interests, FWS recently created an advisory council called the Wildlife and Hunting Heritage Conservation Council, designed to "encourage partnership" between "the public, the sporting conservation community, the shooting and hunting sports industry, [and] wildlife conservation organizations" ("Hunting Heritage"). Instead of protecting and preserving ecosystems on behalf of the majority of citizens, FWS seeks a broader base to support outdated wildlife manipulation on behalf of hunters, including wildlife conservation organizations. No doubt hunting myths will be an important part of cajoling the public into supporting Fish and Wildlife Services ecologically damaging policies.

Hunting myths are also central to the affirmative response of the 109th Congress to the U.S. Hunting Heritage Protection Act—an excellent example of the entrenched, persistent ignorance that surrounds and supports hunting in the United States. For example, members of the 109th Congress erroneously assert that:

(1) recreational hunting is an important and traditional recreational activity. . . .

(2) hunters have been and continue to be among the foremost support-
ers of sound wildlife management and conservation practices in the
United States. . . .

(5) recreational hunting is an essential component of effective wildlife
management. . . .

(7) recreational hunting is an environmentally acceptable activity that
occurs and can be provided for on federal public land without adverse
effects on other uses of the land. ("S. 1522: 109th")

An informed Congress—and an informed public—would offer a very
different response to the Hunting Heritage Protection Act, likely not-
ing that hunting is not a tradition, that hunters (with the help of government wildlife agencies and billions of tax dollars) have done nothing
but serve their own interests *at the expense of ecosystems and wildlife*,
and that hunters invariably have an adverse effect on other public lands
users—even driving other citizens off of public lands in fear of their
lives. All this aside, shouldn't we see red flags when a tiny minority of
citizens tries to seal into law their personal choice of recreation as a
protected heritage?

In general, U.S. citizens have been and continue to be deceived in a
number of ways with regard to hunting. Collectively, we are particularly
uninformed with regard to FWS and the historic link between government conservation agencies and hunting, including ongoing government misappropriations on behalf of hunters and resultant egregious
environmentally damaging policies. Hopefully, you are now aware of
common public misconceptions surrounding hunters and hunting, and

- essential to conservation
- economical
- tradition and/or heritage
- rooted in fair chase"
- merciful (for overpopulated wildlife)
- compassionate (compared with farm-raised meat)
- ecologically sound

(Clker.com)

s3.18 Common Hunting Misconceptions Hold that Hunting is . . .

are therefore prepared to ponder a key question: On what legitimate grounds would any informed, sincere environmentalist support hunting or align with hunter interests?

References

"A Leopold Biography—Part II." *About.com*. Accessed 11 Sept. 2011. <http://forestry.about.com/cs/foresthistory1/a/al_leo_lorb2.htm>.

"About the B & C Club: Boone and Crocket Club since 1987." Accessed 25 May 2012. <http://www.boone-crockett.org/about/about_overview.asp?area=about>.

"Ambush." *M-W.com*. Accessed 10 June 2012. <http://www.merriam-webster.com/dictionary/ambush>.

"Animal Cruelty and Human Violence: A Documented Connection." Humane Society of the United States. Accessed 27 Jan. 2013. <http://www.humanesociety.org/issues/abuse_neglect/qa/cruelty_violence_connection_faq.html>.

"Assassinate." *M-W.com*. Accessed 10 June 2012. <http://www.merriam-webster.com/dictionary/assassinate>.

Bekoff, Marc. *Animals Matter: A Biologist Explains Why We Should Treat Animals with Compassion and Respect*. Boston: Shambhala 2007.

Birdseye, Kari. "Returning to Save the Wolves From the Exterminators." *EarthJustice*, Spring 2014. 12–13.

Carroll, Joseph. "Gun Ownership and Use in America: Women More Likely than Men to Use Guns for Protection." Nov. 22, 2005. Accessed 29 May, 2014. <http://www.gallup.com/poll/20098/gun-ownership-use-america.aspx>

"Caught in the Crosshairs: Effective Immunocontraception Faces Political Fire." *AWI Quarterly* 61.1 (Winter 2012): 22–24.

"Cruel." *M-W.com*. Accessed 26 Jan. 2013. <http://www.merriam-webster.com/dictionary/cruel>.

"Digest of Federal Resource Laws of Interest to the U.S. Fish and Wildlife Service: Federal Aid in Wildlife Restoration Act." U.S. Fish and Wildlife Service. Accessed 1 Nov. 2013. <http://www.fws.gov/laws/lawsdigest/FAWILD.HTML>.

"Environmentalist." *The Free Dictionary*. Accessed 19 Nov. 2013. <http://www.thefreedictionary.com/environmentalist>.

"Excerpts from the Works of Aldo Leopold." Accessed 29 May 2014. <http://www.panojohnson.com/leopold-quotes.html>

Farquhar, Brodie. "Montana Governor's Defiance of Feds Has Few Parallels." *New West* 21 Feb. 2011. Accessed 22 Feb. 2011. <http://www.newwest.net/topic/article/montana_governors_defiance_of_feds_has_few_parallels/C41/L41/>.

"Federal Aid Division—The Pittman-Robertson Federal Aid in Wildlife Restoration Act." U.S. Fish and Wildlife: Southeast Region. Accessed 24 May 2012. <http://www.fws.gov/southeast/federalaid/pittmanrobertson.html>.

Fox, Camilla H. "Carnivore Management in the U.S.: The Need for Reform." *AWI Quarterly* 58.4 (Fall 2009): 20–24.

——. "The Coy Coyote." *AWI Quarterly* 58.1 (Winter 2009): 24–25.

——. "Predators in Peril: The Federal Government's War on Wildlife." *Project Coyote*. Accessed 14 Oct. 2011. <http://www.projectcoyote.org/newsreleases/news_predatorinperil.html>.

Fox, Stephen. *The American Conservation Movement*. Madison: U of Wisconsin P, 1981.

"Frequently Asked Questions about Community Gardens." *Food Share*. Accessed 5 June 2012. <http://www.foodshare.net/files/www/Growing/Community_Garden_FAQ.pdf>.

"Gifford Pinchot (1865–1946)." *The National Parks: America's Best Idea*. PBS. *Historical Figures*. Accessed 29 May, 2014. <http://www.pbs.org/nationalparks/people/historical/2/#pinchot>.

Gross, Ken. "A Tragic Hunting Accident in Maine Kills a Mother in Her Own Backyard and Ends in the Shooter's Acquittal." *People* 5 Nov. 1990. Accessed 15 Feb. 2011. <http://www.people.com/people/archive/article/0,,20113512,00.html>.

Gudorf, Christine E., and James E. Huchingson. *Boundaries: A Casebook in Environmental Ethics*. Washington, DC: Georgetown UP, 2010.

"Guiding Principles." National Wildlife Refuge System: U.S. Fish and Wildlife Service. Accessed 27 Feb. 2012. <http://www.fws.gov/refuges/about/mission.html>.

"Gun Ownership and Use in America." Gallup. 22 Nov. 2005. Accessed 27 May 2012. <http://www.gallup.com/poll/20098/gun-ownership-use-america.aspx>.

Hatfield, Linda. "Report on Bowhunting." Accessed 29 May 2014. <http://animalrightscoalition.com/doc/bowhunting_report.pdf>.

Heffernan, Tim. "Weaponry: The Deer Paradox" *Atlantic* Nov. 2012: 25–26.

"Helping Them All: A Status Report on Wyoming's State Wildlife Action Plan." Wyoming Game and Fish Department. 2010.

"History of the Boone and Crockett Club: The Legacy." Boone and Crocket Club. Accessed 28 Dec. 2012. <http://www.boone-crockett.org/about/about_history.asp?area=about>.

Horkheimer, Max, and Theodor W. Adorno. *Dialectic of Enlightenment: Philosophical Fragments*. Ed. Gunzelin Schmid Noerr. Trans. Edmund Jephcott. Stanford, CA: Stanford UP, 2002.

"How Much on Average Does Deer Hunting Cost? I Have Never Been Hunting Before and Would Like to Start." *Yahoo Answers*. Accessed 5 June 2012. <http://answers.yahoo.com/question/index?qid=20110503174847AA5QUQD>.

Humane Society of the United States. *Learn the Facts About Hunting*. Washington, DC: Humane Society of the United States, n.d.

"Hunters out of the Whole Population in the U.S. What Percentage Actually Participates in Hunting?*?" *Yahoo Answers*. Accessed 5 June 2012. <http://answers.yahoo.com/question/index?qid=20080823102552AAnIkxv>.

"Hunting." U.S. Fish and Wildlife Services. Accessed 1 Feb. 2013. <http://www.fws.gov/hunting/>.

"Hunting Deer." *Your Guide to Hunting on National Wildlife Refuges.* National Wildlife Refuge System. Accessed 24 May 2012. <http://www.fws.gov/refuges/hunting/featured_articles.cfm?heid=2>.

"Hunting Heritage Protection Act Introduced in the Senate." *Brian.Carnell.com.* 16 Aug. 2005. Accessed 29 May 2014. < http://brian.carnell.com/articles/2005/hunting-heritage-protection-act-introduced-in-senate/>.

Kheel, Marti. "The Killing Game: An Ecofeminist Critique of Hunting." *Journal of the Philosophy of Sport* 23 (1996): 30–44.

———. "License to Kill: An Ecofeminist Critique of Hunters' Discourse." *Women and Animals: Feminist Theoretical Explorations.* Ed. Carol Adams and Josephine Donovan. Durham, NC: Duke UP, 1995. 85–125.

———. *Nature Ethics: An Ecofeminist Perspective.* Lanham, MD: Rowman & Littlefield, 2008.

Leopold, Aldo. *A Sand County Almanac.* New York: Oxford UP, 1968.

———. "Review of Young and Goldman, *The Wolves of North America.*" *The River of the Mother of God.* Ed. Susan L. Flader and J. Baird Callicott. Madison: U. of Wisconsin, 1991. 320–22.

———. "The Varmint Question." *Pine Cone* Dec. 1915. Reprinted in Susan L. Flader, *Thinking Like a Mountain: Aldo Leopold and the Evolution of an Ecological Attitude Toward Deer, Wolves, and Forests* (Madison: U of Wisconsin P, 1974), xxvi.

Lewis, Michael S., and Ray Paige. "Summary of Research: Selected Results from a 2006 Survey of Registered Off-Highway Vehicle (OHV) Owners in Montana." Montana Fish, Wildlife, and Parks. Accessed 14 June 2013. <http://fwpiis.mt.gov/content/getItem.aspx?id=19238>.

"The Link between Animal Abuse and Serial Killers." *Pro Death Penalty Resource Community.* Accessed 26 Jan. 2013. <http://off2dr.com/smf/index.php?topic=13413.0>.

Luke, Brian. *Brutal: Manhood and the Exploitation of Animals.* Urbana: U of Illinois P, 2007.

———. "A Critical Analysis of Hunters' Ethics." *Environmental Ethics* 19.1 (1997): 2.

MacInnes, Laura. "U.S. Most Armed Country with 90 Guns per 100 People." *Reuters.* Aug. 28, 2007. Accessed 29 May, 2014. <http://www.reuters.com/article/2007/08/28/us-world-firearms-idUSL2834893820070828>

"Massacre." *M-W.com.* Accessed 10 June 2012. <http://www.merriam-webster.com/dictionary/massacre>.

"Minié Rifle." *Wikipedia.* Accessed 5 June 2012. <http://en.wikipedia.org/wiki/Mini%C3%A9_rifle>.

"Mission Statement." National Wildlife Refuge System: U.S. Fish and Wildlife Service. Accessed 26 Jan. 2013. <http://www.fws.gov/refuges/about/mission.html>.

Moss, Doug. "E Word: Operation Prairie Storm." *E Magazine* July/Aug. 2004: 6.

Napier, William Francis Patrick. *The History of General Sir Charles Napier's Administration of Scinde.* London: Chapman & Hall, 1851.

"2006 National Survey of Fishing, Hunting, and Wildlife-Associated Recreation." *U.S. Fish and Wildlife Services*. Accessed 30 May 2014. <http://www.census.gov/prod/2008pubs/fhw06-nat.pdf>.

"2011 National Survey of Fishing, Hunting, and Wildlife-Associated Recreation." *U.S. Fish and Wildlife Services*. Accessed 30 May 2014. <http://www.census.gov/prod/2012pubs/fhw11-nat.pdf>.

"Petition to Reform Wildlife Services." *AWI Quarterly*, Winter 2014, 63:1. 25.

"Pittman-Robertson Wildlife Restoration." *FY 2013 Budget Justification: Wildlife Restoration*. U.S. Fish and Wildlife Services. 1–12. Accessed 12 Jan. 2013. <http://www.fws.gov/budget/2013/PDF%20Files%20FY%202013%20Greenbook/24.%20Wildlife%20Restoration.pdf>.

"Police: Hunter Accidently Shoots Man, 13-Year-Old Boy; Man, Son Injured In Shooting." *WXII12.com*. 31 December 2010. Accessed 20 Dec. 2012. <http://www.wxii12.com/r/26332551/detail.html>.

"Public Land." *Wikipedia*. Accessed 12 Jan. 2013. <http://en.wikipedia.org/wiki/Public_land>.

"Quotations about Animal Rights." *The Quote Garden*. Accessed 27 May 2012. <http://www.quotegarden.com/a-rights.html>.

"Quotes about Hunting." *Goodreads*. Accessed 12 June 2012. <http://www.goodreads.com/quotes/tag/hunting>.

"Reducing Wounding Losses." South Dakota Game, Fish and Parks. Accessed 13 Jan. 2013 <http://gfp.sd.gov/hunting/waterfowl/wounding-losses.aspx>.

Reed, Len. "Wildlife Watching Surpasses Hunting and Fishing." *Oregonian*. 16 Nov. 2008. Accessed 12 Mar. 2011. <http://www.oregonlive.com/news/index.ssf/2008/11/wildlife_watching_surpasses_hu.html>.

Regan, Tom. *Defending Animal Rights*. Urbana: U of Illinois P, 2001.

"Refuge." *M-W.com*. Accessed 10 June 2012. <http://www.merriam-webster.com/dictionary/refuge>.

"Rifle." *Wikipedia*. Accessed 5 June 2012. <http://en.wikipedia.org/wiki/Rifle>.

Robertson, Jim. *Exposing the Big Game: Living Targets of a Dying Sport*. Winchester, UK: Earth Books, 2012.

Roney, Marty. "Girls, Women Aim for Hunting." *USA Today* 31 Mar. 2008. Accessed 12 May 2011. <http://www.usatoday.com/news/nation/2008-03-31-hunting-girls-women_N.htm>.

Scully, Matthew. *Dominion: The Power of Man, the Suffering of Animals, and the Call to Mercy*. New York: St. Martin's, 2002.

"Smith and Wesson: Handguns." Accessed 9 May 2011. <http://www.smith-wesson.com/webapp/wcs/stores/servlet/Category3_750001_750051_757751_-1_Y#>.

"Statistical Data Tables for Lands under Control of the Fish & Wildlife Service (as of 9/30/2013)." Accessed 29 May 2014. <http://www.fws.gov/refuges/realty/archives/pdf/2013_Annual_Report_of_LandsDataTables.pdf>

"S. 1522: 109th Congress, First Session." 28 July 2005. Accessed 11 June 2012. <http://www.funoutdoors.com/files/S%201522%20Chambliss.pdf>.

"S.1522 Hunting Heritage Protection Act." Born Free USA: Federal Legislation. Accessed 11 June 2012. <http://www.bornfreeusa.org/legislation.php?p=211&more=1>.

"Tradition." *Dictionary.com.* Accessed 12 June 2012. <http://dictionary.reference.com/browse/tradition>.

2006 National Survey of Fishing, Hunting, and Wildlife-Associated Recreation. U.S. Fish and Wildlife Service. Accessed 24 May 2012. <http://digitalmedia.fws.gov/cdm/ref/collection/document/id/21>.

Watson, Paul. "Foreword." *Exposing the Big Game: Living Targets of a Dying Sport.* Jim Robertson. Winchester, UK: Earth Books, 2012. 1–9.

Webster's Encyclopedic Unabridged Dictionary of the English Language. New York: Gramercy, 1996.

"Why Sport Hunting Is Cruel and Unnecessary." PETA Wildlife Factsheet. Accessed 8 Feb. 2011. <http://www.peta.org/issues/wildlife/wildlife-factsheets/sport-hunting-cruel-unnecessary/>.

"PA: Woman Shot by Hunter in Lancaster County." *Committee to Abolish Sport Hunting.* 19 Jan. 2011. Accessed 20 Feb. 2011. Accessed 20 May 2014. <http://www.all-creatures.org/cash/taah-sh-20110202-9.html>.

Yarrow, Greg. "The Basics of Population Dynamics." Clemson Cooperative Extension: Extension Forestry and Natural Resources. May 2009. Accessed 1 May 2013. <http://www.clemson.edu/extension/natural_resources/wildlife/publications/fs29_population_dynamics.html>.

"Your Guide to Hunting on National Wildlife Refuges." National Wildlife Refuge System. Accessed 27 Feb. 2012. <http://www.fws.gov/refuges/hunting/>.

Conclusion

EACH CHAPTER OF *Eating Earth* focuses on environmental problems that stem from our willingness to consume animal products—environmental problems that stem from animal agriculture, fisheries, and hunting policies and practices. The final chapter also exposes relevant history and myths that explain why the vast majority of U.S. citizens continues to accept sport hunting despite the environmental problems that stem from this deadly pass time. Animal products, whether organic or local, whether hunted or purchased, whether chicken or fish or yogurt, harm the environment.

As this book draws to a close, it is important to at least mention the larger picture with regard to ethics and dietary choice: There are a handful of other critical reasons to move decisively toward a plant-based diet, all of which are interconnected. I remember the five reasons for choosing a plant-based diet through a mnemonic using the Italian word for love, AMORE.

In this acronym, "A" represents what is likely the most common reason for choosing a vegan diet—animals. In choosing to kill or buy body parts, mammary secretions, and eggs from other animals, we support the exploitation and slaughter of living, breathing, sentient beings, who would prefer to live out their natural lives peacefully in their own communities. In the U.S., ten billion farmed animals are denied pretty much every natural behavior, without regard to their sufferings, only to be shipped to their deaths when they are adolescents—all for the sake of eggs, milk, and various "meats." The long-term suffering endured by farmed animals—especially female farmed animals in the egg and dairy industries—is truly unconscionable. Cruel practices are *always* unearthed when undercover investigators penetrate the increasingly thick walls that conceal common animal agriculture practices. If you are unaware of the stunted lives and premature deaths forced on farmed animals around the world, please explore footage taken by undercover activists[1], starting with the excellent YouTube clip, "If Slaughterhouses Had Glass Walls."

"M" represents the many critical medical reasons for rejecting an omnivore's diet. Each of the leading health-related causes for premature death in the U.S. is linked with dietary choice—one is indirectly linked, and the other four are *directly* linked with the consumption of animal products:

1. Heart disease
2. Cancer
3. Respiratory problems (indirect)
4. Stroke
5. Diabetes

How many people living on fresh fruits and tofu suffer a heart attack when they are in their forties? How often does a steady diet of whole grains and leafy greens lead to colon cancer? How many strokes are brought on by green smoothies and sprout sandwiches? How often do bean salads and barley soup cause obesity?

Health is not just a matter of our own quest to live long and well. Our health affects others, especially those who care about us, those who depend on us. . . and those who ultimately pick up the tab for expensive medical procedures. How are partners and children affected if we experience a paralyzing stroke, or if we suffer from the many expected medical problems associated with diabetes, including premature death? As a community, how many triple bypass surgeries caused by dietary choice are we willing to pay for? How might we prefer to invest this money?

Much suffering and many medical bills would be altogether avoided if, as a community, we shifted to a plant-based diet—think of what communities and nations could do with those billions of dollars! There is no biological need to consume animal products. There is ample evidence (for example, T. Colin Campbell's work published in *The China Study*) that optimal health is achieved with a diet of greens, grains, and fruits.

1. VIVA! USA (http://www.youtube.com/watch?v=QFomoIUaZ-k), PETA (http://www.petatv.com/), HSUS (http://video.hsus.org/), PCRM (http://www.pcrm.org/resources/), Farm Sanctuary (http://www.farmsanctuary.org/mediacenter/videos.html), and Vegan Outreach (http://www.veganoutreach.org/whyvegan/animals.html).

...Which Leading Killers Stem from Dietary Choice?

Primary Killers (deaths per year)		Meat, Dairy, Eggs	Plant-Based Diet
Heart Disease (697,000)		Yes	No
Cancer (557,300)		Yes	No
Stroke (162,700)		Yes	No
Respiratory Disease (124,800)		Yes*	No
Diabetes (74,000)		Yes	No
Obesity		Yes	No

(clker.com)

*Thanks to animal ag's mighty contribution to air pollution, respiratory diseases are also linked (indirectly) to diet.

FIGURE C.1 What Is the Healthiest Way to Feed a Family?

"O" stands for oppressed people. Across nearly five decades, eco-feminists have explored the interconnected nature of oppressions, unearthing and describing a number of ways that animal products further oppress the earth's most downtrodden peoples. Those likely to be oppressed by our omnivorous diet include the world's hungry: as noted in chapter 1, 70 percent of U.S. grain and 60 percent of EU grain is fed to farmed animals while people die every second of every day for want of sustenance. Ecofeminists have also linked animal agriculture with heterosexism (emphasis on breeding, for example), racism and classism (as with subsidized junk food eroding the health of poorer, inner-city communities), ableism (devaluing those who are dependent, whether cows or human beings, for instance), and sexism (exploitation of female reproductive systems for milk and eggs). Choosing to eat animal products supports and maintains these systems of oppression (see Carol Adams, ed., *The Pornography of Meat*; Breeze Harper, ed., *Sistah Vegan*; and Lisa Kemmerer, ed., *Sister Species: Women, Animals, and Social Justice*).

"R" stands for "religion." Religions and spiritual philosophies around the world discourage people from causing suffering and from destroying life and the natural environment. The world's great religions encourage humanity to be respectful of life and the earth, to be respectful toward our bodies, and to work to alleviate hunger. Religious moral codes require respect for creation/

living beings and compassion for all—including nonhuman animals, laborers, and the poor (See Lisa Kemmerer, *Animals and World Religions*).

Of course "E" stands for Environment (or Earth). Placed in a larger context, protecting the planet—atmosphere, forests, freshwater, soils, wildlife, and ecosystems—is just one of five compelling, interconnected reasons to rethink the consumption of animal products. *Eating Earth* is therefore written for anyone who cares about animals, human health and the rising costs of medical care, dismantling systems of oppression (or simply feeding the hungry), fulfilling basic religious commitments, and/or protecting and healing the environment.

As noted, this book is written for environmentalists—all environmentalists, but especially mainstream environmentalists. Nonetheless, concerns beyond the environment that are associated with an omnivorous diet are evident throughout this book, including animal lives and wellbeing, human health, and oppressed peoples. For example, chapter 1 mentions the social justice concerns entailed in feeding corn and soy to cattle, pigs, and chickens while other human beings suffer from malnutrition and starvation; chapter 2 mentions human health with regard to mercury in fish; and chapters 2 and 3 discuss suffering in fish and hunted animals. In light of such pressing, interlocking concerns surrounding our tendency toward omnivory, it is counterintuitive and perhaps even unjust and irresponsible to focus exclusively on only one of the interconnected problems associated with this particular dietary choice. The other critical concerns deserve to at least be mentioned, shedding light on the somewhat daunting extent of the damage caused by omnivorous dietary choices.

On reflection, in light of the five interconnected reasons stated above, a single-minded focus on the environment does not do justice to the importance of choosing a vegan diet. (For a great short video on these interlocking reasons, see "Why Vegan?" by EVOLVE Campaigns, http://www.youtube.com/watch?v=8GrbYVsK7vs.) Not only environmentalists but anyone who cares about human beings—health and the world's poor—or is concerned about the sufferings of pigs and fish and pheasants, or who is committed to one of the world's great religions, ought to choose a vegan diet.

Environment, Diet, and Choice

Although there are at least five urgent reasons to shift to a plant-based diet, no reason is likely to be as compelling to an environmentalist as those highlighted in this book. I know this from talking to the Irishman mentioned in the introduction: Though I had done my best to explain the sufferings of farmed animals, the religious importance of looking after less powerful citizens of the earth, and other reasons to shift to a vegan diet, my words failed, year after year, to inspire a change of diet. Yet, when confronted with *just one* of many environmental problems brought on by choosing an omnivorous diet—in this case the ratio of the mass of wild birds to that of chickens in the U.K. (1:104)—he grasped how a taste for egg salad sandwiches and chicken noodle soup warps the natural balance of life in the U.K. He wrote of this moment: "Truly, for me the realization was like scales falling from my eyes." In that moment he was forced to ask himself how the taste of chicken or eggs could possibly be worth such hefty environmental consequences.

In our day-to-day lives we often act without thinking, behave without thoughtful intent, and live without conviction. This is nowhere more evident than in our eating habits. In pretty much any nation it is easy to travel from the cradle to the grave without ever thinking about what we put into our mouths—without any awareness of the effects of dietary choice on the earth, on animals, on our health, and on disempowered people—without any awareness that *inasmuch as we choose what we eat, diet is a matter of utmost moral significance.*

Based on dietary differences, due to consumer power, certain foods are common in some places but difficult (if not impossible) to find in other places. British citizens tend to recoil at the thought of eating developing chicken eggs—a delicacy in the Philippians. Botswana's people often find caterpillars to be quite tasty, while Canadians rarely choose to dine on insects. India has been overwhelmingly vegetarian for centuries; China has been without dairy products for yet longer (and incidences of osteoporosis are very rare in China). Consequently, compared with more traditional foods, it is difficult to find ready-to-hatch eggs in British food markets, caterpillars in Canadian food markets, beef in Indian food markets, and cow's milk in Chinese food markets. This demonstrates the power of the consumer—especially in capitalistic nations. *By shifting to a vegan diet we simultaneously shift food industries, altering which foods will be available in local markets, restaurants, and dining halls.* The complete and utter dependence of animal agriculture and fisheries on our dietary choice—on consumers—is both empowering and liberating: each of these industries/practices has come into existence because of human consumption patterns, and their continued existence depends wholly on a continuation of these particular patterns. If we shift to a plant-based diet, animal industries and fisheries will be replaced with more earth-friendly alternatives. Our actions as consumers are our surest vote.

For anyone reading this book, diet is almost surely a matter of ethics—a matter of choice. We can choose to replace eggs, dairy, and flesh with plant-based foods, with a variety of tasty alternatives (such as Daiya cheeses, Purely Decadent ice creams, Yves spicy Italian sausages, and WholeSoy yogurts). The transition from omnivore to herbivore is especially easy if we connect with other vegans, sampling their favorite foods and recipes.

Eating animals is eating earth. Either the foods we choose are linked with high greenhouse gas emissions, or they are not. Either the foods we serve for lunch cause considerable ecosystem disruption, or they do not. Either the restaurants that we frequent are environmentally friendly, or they are not. Either we vote for animal industries, fisheries, and hunting with our food dollars, or we do not. How can we legitimately consider ourselves to be environmentalists if concern for the planet does not guide our most basic daily choices? On what rational grounds would sincere, informed environmentalists *choose* to support the most environmentally damaging industries and practices with their consumer dollars?

Food choices are the number one determinant of an individual's environmental footprint. While many environmentalists advocate for smaller cars, recycling, fluorescent lights, and shorter showers, these changes can bring only very minimal planetary benefit compared with adopting a plant-based diet. Environmentalists cannot reasonably ask others to change consumer habits (whether taking public transport, shopping with cloth bags, or letting lawns go brown in the summer) if unwilling to make the most fundamental change necessary on behalf of the planet in their own daily lives—dietary change.

You are now among those who understand the serious environmental implications of dietary choice. As a consumer—in supermarkets and convenience stores, in take-away stands and restaurants, in sporting goods stores and wildlife refuges—you either vote for or against the earth with your pocketbook. Every time you eat, you either choose to chew and digest high planetary impact animal products or you do not. Please vote for the earth with your pocketbook and your teeth. Committing to a plant-based diet is the most important decision you can make on behalf of the environment.

Index